WORLD BANK TECHNICAL PAPER NO. 389

I0120162

Planning the Management, Operations, and Maintenance of Irrigation and Drainage Systems

A Guide for the Preparation of Strategies and Manuals

International Commission on Irrigation and Drainage (ICID)

The World Bank
Washington, D.C.

Technical Papers are published to communicate the results of the Bank's work to the development community with the least possible delay. The typescript of this paper therefore has not been prepared in accordance with the procedures appropriate to formal printed texts, and the World Bank accepts no responsibility for errors. Some sources cited in this paper may be informal documents that are not readily available.

ISSN: 0253-7494

The photograph on the cover was taken by Mr. P. Rousset from La Societe du Canal de Provence, France, in the Majalgaon Pilot Modernization Project, Maharashtra State, India.

Library of Congress Cataloging-in-Publication Data

Planning the management, operation, and maintenance of irrigation and
    drainage systems : a guide for the preparation of strategies and
    manuals / International Commission on Irrigation and Drainage. —
[2nd ed.]
        p.   cm. — (World Bank technical paper ; no. 389)
    Includes bibliographical references (p.).
    ISBN 0-8213-4067-0
    1. Irrigation—Management.   2. Drainage—Management.
I. International Commission on Irrigation and Drainage.   II. Series.
TC812.P53   1997
627'.52'068—dc21   97-37307
                                                         CIP

TABLE OF CONTENTS

LIST OF FIGURES[1]

1/These figures are the author's calculations, unless otherwise noted.

FOREWORD

The first edition of this Technical Paper was published in June 1989 as a joint effort of the World Bank and the International Commission on Irrigation and Drainage (ICID). Those organizations are pleased to join together again in order to publish a second edition of the paper, which provides the central reference for the preparation of Plans of Operation and Maintenance (POMs) and includes a comprehensive list of the issues which should be addressed in a POM.

In the foreword to the 1989 edition our predecessors wrote:

Good management, efficient operation and well-executed maintenance of irrigation and drainage systems are essential to the success and sustainability of irrigated agriculture. They result in better performance, better yields from crops, and sustained production.

Unfortunately, management operation and maintenance are often poorly carried out. The main reason has been generally attributed to inadequate finance. While it is clear that adequate finance is a prerequisite, experience has shown that weaknesses in institutional, technical, and managerial aspects of an irrigation and drainage organization are also important factors that constrain good system performance.

With the growing competition both for water and finance, the emphasis on good management and efficient operation and maintenance has grown ever more important. This has increased awareness of the need for more performance-orientated use of water and financial resources, wherein the organizations responsible for the operation and maintenance of large systems need to work ever more closely with farmers to achieve the level of service which best meets the overall objective of irrigated agriculture, namely sustained improvement in agriculture production.

The original edition of this Technical Paper was the culmination of several years of effort by the ICID Working Group on Operation, Maintenance and Management of Irrigation, Drainage and Flood Control Projects. The same working group has been involved in the production of the second edition with a new Introduction giving a greater emphasis on the need for the management of operation and maintenance to be approached according to the principles developed for the UN Conference in Environment and Development (UNCED), held in Rio de Janeiro in 1992 and the International Conference on Water and Environment (ICWE) in Dublin which preceded it. The ICID has responded positively to the challenge which the achievement of these principles presents. Not least in the operation and maintenance of irrigation and drainage, which in the spirit of Rio de Janeiro and Dublin, requires management to be much more aware of the ultimate objectives and the need for cooperative linkage with all the various interests concerned, and in particular, a sense of service to the farmers and other beneficiaries.

The need for operation and maintenance to be carefully formulated during the design stage of a project was one of the conclusions of Question 44, at the 15th ICID Congress, in the Hague, 1993. In Question 45 at the same Congress, the call was for more decentralized management with less of the O&M burden being presumed to be publicly funded and for O&M organizations themselves to be more focused and devolve more responsibility and authority to water users. The implications of this on O&M organizations is reflected in the introduction to the second edition.

Dr. M. A. Chitale
Secretary General
International Commission on Irrigation and Drainage
48 Nyaya Marg, Chanakyapuri
NEW DELHI, 110021, India

Aly Shady
President
International Commission on
Irrigation and Drainage

Herve Plusquellec
Irrigation Advisor
The World Bank

Alex McCalla
Director, Rural Development Department

PREFACE

The first edition of this guide for the preparation of strategies and manuals in the operation and maintenance of irrigation and drainage projects, was produced by the ICID's Working Group on Operation, Maintenance and Management, under the chairmanship of Mr. J M Schaack. This is the culmination of several years work beginning in 1983 and most of the material produced in the first edition has been carried forward into this new edition, including the case studies provided by the Narmada Sagar in India (*Appendix 2.1*) and the organization of operation and maintenance exampled by the Goulburn-Murray Irrigation Authority in Australia (*Annex 2.2*).

The actual production of the first edition was largely due to the support of the World Bank and the Senior Irrigation Advisor, Mr. Guy Le Moigne and the ICID/World Bank editorial group chaired by Professor D. J. Constable of Australia. This group comprised many well known names in irrigation and drainage, and because so much of their work lives on in this second edition it is appropriate to acknowledge their contribution again.

The members of the editorial group comprised:

> Professor D. J. Constable (Australia), Chairman
> Mr. H. Frederiksen (World Bank, Washington, D.C.)
> Mr. H. M. Hill (Canadian National Committee, ICID)
> Mr. G. Le Moigne (World Bank, Washington, D.C.)
> Mr. W. J. Ochs (World Bank, Washington, D.C.)
> Mr. H. Plusquellec (World Bank, Washington, D.C.)
> Mr. W.R. Rangeley (United Kingdom), President Honoraire, ICID
> Mr. J. Sagardoy (FAO, Rome)
> Mr. J. M. Schaack (USA National Committee, ICID)

It was Professor Constable who saw a need for a substantial revision to the introduction to the second edition to further develop the planning framework for POMs. This was needed to acknowledge the principles developed at the International Conference on Water and the Environment (ICWE) and outlined in the Dublin Statement, which became the basis for Chapter 18 of Agenda 21 of the Earth Summit held in Rio de Janeiro in 1992. This saw the need for management of irrigation and drainage to be based on a participatory approach, linking social and economic development with the protection of natural eco-systems. Whilst it is a basic right of all human beings to have access to water, it needs to be recognized that water has an economic value and needs to be managed accordingly.

The participatory approach and the need to recognize the economic value of water has brought about fundamental changes in the way people view the sustainability of irrigation and drainage systems. With much of the world's water resources now allocated to existing uses, the ability of irrigation to feed a growing world population is largely dependent on increasing the effectiveness of existing irrigation development. The process of operation and maintenance is therefore giving new emphasis to the management of water, financial and human resources.

Fundamentally this requires irrigation agencies to deliver a level of service which benefits the users and means the cost of service is affordable, and implicitly makes better use of the resources the service is consuming. In short we are beginning to see irrigation treated much more as a business in which its assets are used in such a way that it will generate sufficient revenue to sustain the business, and renew and modernize those assets to obtain greater value from their use of water and land, the two fundamental resources which in future are going to become more limited.

These thoughts are implanted in the second edition by a revision of the Introduction and particularly Section B on the planning framework, which is largely the work of Professor Constable. The changes here and in the rest of the document have been put together with the assistance of Mr. H. Plusquellec of the World Bank who arranged for the original manuscript to be completely retyped and the small editorial team, drawn from the present membership of the ICID Working Group on Operation, Maintenance and Management, who put together the final manuscript.

The members of the editorial group for the second edition comprised:

Mr. P S Lee (UK), Chairman
Dr. H M Malano (Australia)
Mr. E Caligiuri (Canada)

ABSTRACT

This technical paper has been prepared in the new second edition to provide a guide for organizations responsible for operation and maintenance of irrigation and drainage systems, to develop strategies and plans for proper and effective operation and maintenance.

The format of the main part of the paper provides the basis for the preparation of a plan of operation and maintenance (POM), with a comprehensive list of the issues which need to be addressed in such a plan and in more detailed operation and maintenance manuals. The management aspects of the process are emphasized.

The new introduction of the second edition highlights the need for clear management objectives and a coordinated cooperative linkage with other organizations and in particular, setting a specific level of service which will provide a measure of performance and justify the cost of what the organization is doing.

INTRODUCTION

This guide has been prepared as a reference document for organizations that are responsible for the operation and maintenance (O&M) of irrigation and drainage systems.

The aim of the guide is to assist such organizations in developing their strategies and in preparing plans for O&M (POM).

Within this general objective, the guide provides a basis for the preparation of O&M Manuals which give the essential operating instructions to managers in the field and other O&M staff, and which form part of the POM.

There is a multitude of activities, programs and functions which must be effectively planned, executed and coordinated if the organization is to discharge those responsibilities assigned to it.

The more important of these responsibilities are obligations to operate and maintain the facilities to meet the project capability as designed and constructed, but also to meet the expectations of its water consumers and their dependent communities on an ongoing basis.

The guide in itself is not intended to give specific instructions and directions for all of those activities and programs. Rather, its purpose is to provide a comprehensive list of all issues to be addressed, together with a listing of published material and working papers which will assist in the formulation of a specific POM and its associated manuals.

CONTEXT AND SCOPE OF THIS GUIDE

There is a multitude of organizational arrangements represented by the collective experience of the many countries of the world in which irrigated agriculture is practiced, and a variety of water distribution systems. Within these arrangements, there are many different management systems by which responsibilities for operation and maintenance of a particular irrigation network may be assigned to one or more agencies, or to a unit or units of a particular agency.

These organizational arrangements may have evolved from custom or be a result of a specific decision to establish institutions to achieve particular goals relative to improving performance in the irrigated agriculture sector.

The organizations may:

- range from Government institutions through to privately-managed Associations or Cooperatives of water users with little Government involvement;

- comprise multi-purpose, multi-functional bodies with a wide range of responsibilities, or special purpose organizations with water management in one project, as a primary role;

- be financed totally by Government or private funds or by a combination of the two.

The water distribution system being managed may be:

- long-established "traditional" schemes;

- existing schemes which have been rehabilitated or modernized;

- newly-commissioned projects;

- "run-of-the-river" diversions that involve large headwork storages and complex distribution multi-purpose systems.

A more comprehensive discussion of these alternative arrangements is set out in the publication "Organization, Operation and Maintenance of Irrigation Schemes" by the Food and Agriculture Organization (Ref. 55). Some examples are shown in Annex 2.

In more recent years, increasing attention is being given to greater devolution of responsibility to farmers or water user groups for operation and maintenance of the farm distribution systems ("tertiary systems"), with a corresponding thrust for a stronger managerial and more commercially orientated approach to sustainable development of the water resources and provision of services at the river basin and river system level. Two questionnaires carried out by ICID in 1993/94 and 1994/95 provide background information on the arrangements for these management functions and costs in a number of countries (Ref. 85).

It is against a background of such diversity in management systems that these guidelines have been developed.

The general approach has been to set out essential principles which should guide the development of an effective Plan for Operation and Maintenance (POM) and a list of references which will assist in providing the detail for the POM specific to a particular system.

The guide includes by necessity, some brief and generalized discussion of Institutional Planning and Management arrangements, as a necessary framework within which effective operation and maintenance activities will be performed.

In this context, the guide has been directed primarily to public sector irrigation schemes, because:

the management, financial and personnel processes are frequently less flexible than those for a private scheme, because of the need to conform with central Government administrative requirements.

- public sector schemes are usually large in extent compared to private schemes, and mostly involve a wider range of issues to be comprehended by the managers.

However, many of the specific procedures listed, particularly those dealing with operation and maintenance of the physical elements of the system, are directly applicable to private sector projects.

PLANNING FRAMEWORK

Plan of Operations and Maintenance (POM)

One of the key objectives in the management of an irrigation and drainage system is to provide levels of service as agreed with Government/Project Manager and consumers at minimum achievable cost.

To meet that objective, and assure the ongoing integrity of the facilities embodied in an irrigation project, calls for management skills of a high order. Those skills are required to coordinate effectively the elements of staff, equipment, and the physical and financial resources involved in the project.

A Plan for Operations and Maintenance (POM) is required to establish a strategy for the achievement of these objectives. The POM will form one of the set of management plans with others dealing with such issues as human resources (manpower), finance and development as described in paragraph (b) below.

The POM is a permanent set of documents and instructions, organization charts, work programs and schedules, updated when changes are made, so that it comprises a complete statement for reference and guidance at every level in the project organization.

A major component of a POM will be a Manual comprising a number of sections dealing with different activities and functions. For larger projects, it may have many sections and require several volumes.

Institutional Planning and Management

National Planning Context. The UN Conference on Environment and Development (UNCED), held in Rio de Janeiro in June 1992, made a major step forward in outlining a framework for sustainable development into the 21st century, recognizing that development and environment sustainability are inseparably linked. The deliberations of UNCED drew heavily on the conclusions of the International Conference on Water and the Environment (ICWE) as outlined in the Dublin Statement. A major conclusion of ICWE was that:

"Scarcity and misuse of freshwater pose a serious and growing threat to sustainable development and protection of the environment. Human health and welfare, food, security, industrial development and the ecosystems on which they depend, are all at risk, unless water and land resources are managed more effectively in the present decade and beyond than they have been in the past." (References 82, 83].

One of the central guiding principles in addressing these issues is that:

"Water Management should be approached in a comprehensive inter-sectoral manner, integrating the whole set of policy, institutional, economic, financial, technical, environmental and social dimensions, so as to plan, develop and operate water systems in a sustainable manner." (Ref. 84).

In many countries, water for irrigated agriculture accounts for more than 80 percent of the water extracted from river systems. Therefore, effective management of irrigation systems is a key issue in moving towards sustainable management resources. For an irrigation agency, "sustainable management" must embrace all three elements of the Irrigated Agriculture Production System:

- natural resources of land and water and the associated ecosystems in which the project is located;

- irrigation and drainage infrastructure;

- farm production system.

[Note: The natural resources involved extend beyond the project area - activities within the project area in many cases produce effects to land and water (including groundwater) outside the project lands.]

The development of overall Government policies, objectives and strategies is the outcome of a series of dynamic interactions. Some Government objectives require long periods for their achievement, e.g. self-sufficiency in food production.

However, at any particular time interval there will be a set of specific issues which will need to be addressed in policy and strategy development. The relative priority and criticality of some issues will change over time.

Policies in the Agriculture Sector, for example, can be influenced by international events and pressures, variable climatic conditions (e.g. droughts, floods) as well as by internal economic and social changes.

A particular agency must be able to develop a capacity to continually monitor its operating environment. It will need to identify relevant issues in the contemporary overall social, political and economic context, and it will need to address these specifically in its planning and management processes.

An emerging issue for most agencies, for example, is for them to be able to respond to concerns for longer term environmental balance in managing development processes.

The dynamic interactions which might influence the development of the planning and management objectives for an individual irrigation agency are indicated in Figure 1.

Planning Linkages and Management Interfaces. The internal planning and management processes should be developed so as to effectively manage the interfaces of the agency with its "externalities." Referring to Figure 1, there are four key areas:

- with minister/government

- other responsible agencies, e.g.:
    - Economic Planning
    - Agriculture
    - Natural Resources and Environmental Agencies
    - Social Development and Health
      Communications and Transport
- water-users/farmers

- community

The most obvious and recognizable management interface is between Minister/Government and the irrigation agency. This interface requires the agency to:

- develop plans and programs in accordance with national objectives;

- be accountable to the community for their effective implementation and management.

However, the most critical interface in the context of project performance is that between the project staff responsible for the management, operation and maintenance of the irrigation and drainage systems, and the farmers who use the services provided for irrigated agricultural production on their lands.

The failure in the project design to recognize the existence of this interface, and the associated failure to provide for positive and effective management control of transaction across this interface including measurement of discharges is the most frequent cause of ineffective and inefficient project performance.

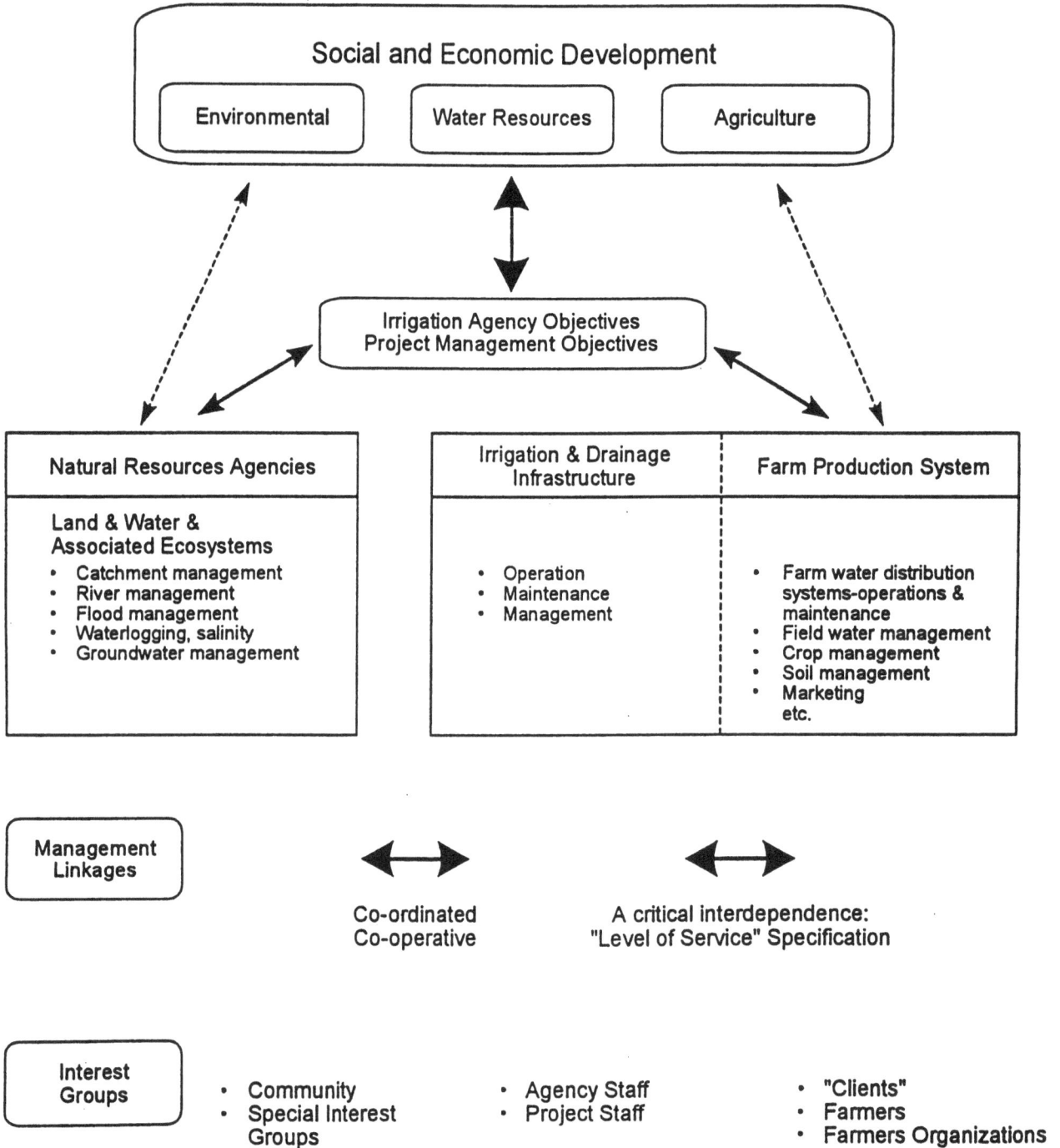

Figure 1   Sustainable Development & Management:
Management Linkages

In such cases, in operation of the system after its construction, many informal arrangements develop, with limited or unclear accountability of the system operators, leading to sub-optimal performance of both the irrigation system and the farm production systems.

Where holdings are large, the interface may occur at the point of supply to individual farms. Where holdings are small, the interface may be at the point of supply to an aggregation of farms, a village or cooperatively farmed lands. In these cases, the development of a coordinating organization downstream of the supply point involving all farm managers within the supply area, is an essential prerequisite to:

- arrange the water distribution to individual farms;

- carry out the maintenance on the farm distribution system;

- identify a contact person for management liaison with the project operations staff.

Significance of "Level of Service" Specification. The "Level of Service" specification is a document which sets out clearly and concisely the nature and extent of services to be provided to farmers at the relevant supply point. The purpose of the document is to provide:

- a basis for design of the canal system;

- a focus for development of the project management objectives and measures for performance evaluation;

- information to farmers of their entitlement to a supply of water;

- a basis for continual review and improvement of canal operation.

Assignments of Water & Entitlements. Assignments of water are generally defined as allocations from the river basin resources sanctioned by Government for utilization within a specific project area. The cumulative assignments within a river basin system should not exceed the safe yield of the particular system, having regard to the levels of security and priorities identified in the overall development planning. These cumulative assignments provide the basis for developing the river regulating and storage release rules.

Within a particular project area, the assignment of water to the project forms the basis of determining the entitlements to water of individual water users.

In projects where water is limiting, i.e. there is not sufficient water to fully irrigate the total area of commanded lands, individual entitlement, representing an individual farmer's share of available water, assumes great significance and value.

As the farming economy develops and matures, the dynamic processes associated with optimizing farm production and income will lead to the development of mechanisms for transfer or leasing of entitlement between farmers on a permanent or temporary basis.

Even though there may be no formal identification of individual water rights at the commencement of a project, the benefits conferred from the "Levels of Service" provisions generally acquire status as a "right" over time.

In most maturing systems, the necessity to define these "rights" in more specific terms arises. In irrigation supply there are two elements embodied in the concept of supply:

- access or title to the water itself, expressed as a right to a specified volume on an annual basis, and in some countries referred to as a water right (constituting a share of the resource);

- the conveyance and distribution of water to farm boundaries, in accordance with agreed rates of supply and delivery periods, either with or without limits to the total volume of supply (constituting a share of available canal delivery capacity).

Obviously, individual entitlements can only be determined in the context of the project assignment of water, having regard to:

- the available capacity of the canal system;

- the canal distribution efficiency, expressed as the percentage of water diverted at the canal intake which can be delivered at the supply point.

The initial determination of entitlement should, therefore, be based on actual operational performance.

As efficiency of operation improves, such improvement opens up options for the further allocation of entitlement from the extra availability of water within the project area as follows:

- increase in existing individual entitlement;

- supply to additional areas within the project;

- a combination of the above;

- transfer of rights to alternative uses.

Objectives for System Management. The three primary objectives for system management include:

- to operate the system to deliver services to farmers in accordance with the Level of Service obligations;

- to maintain the system infrastructure in perpetuity to satisfactory operational standards;

- to manage the system at minimum achievable costs. The linkages between these objectives are displayed in Figure 2.

Financial Management Implications. The requirement for sustainable management will require the development of asset maintenance and management strategies aimed at preventing loss of service capability which would affect the ability to deliver the level of service obligations (*see Figure 2*). The important task of the system managers is to identify the future cash flow requirements to ensure the sustainability of the system, and quantify the effects of under-provision of these cash flows. Even in well-managed systems, such under-provision will lead to a shortening of asset life and inability to maintain supply obligations, with consequential economic loss to farmers and the associated community dependent upon the agricultural production. In the longer term, system owners (Government or farmers as the case may be) will be required to incur substantial financial commitments for restoration of the assets earlier than necessary.

Note that significant improvements in Level of Service provided by the system may involve substantial additional investment. The strategic management processes should include continual review of options for improving Level of Service by optimizing the potential for improved performance of both the canal supply system and the farm production system. Any new investment will be required to satisfy the economic evaluation criteria set by the Government and funding agencies.

Financial Linkages. The need to move towards sustainable systems in the light of the limitations in available land and water resources, and in available financial resources for new and improved irrigation systems as well as for maintenance of existing facilities, will place additional demands on the system managers.

In cases where the operation and maintenance of public sector irrigation projects is carried out by Government Irrigation Departments and funded from central treasuries as an element in government budgets, the relationship between the cost of the service and the capacity or willingness to pay that cost is often obscure. For financially autonomous organizations, i.e. those financed directly (in whole or in part) with funds provided by water users through charges or fees, the relationship is more sharply focused (*see Figure 3*).

However, the adoption of a business-like or "commercial" approach to management of system assets and service delivery, in which giving "value for money" is a primary thrust, is equally important for both.

Figure 2   System Management Objectives

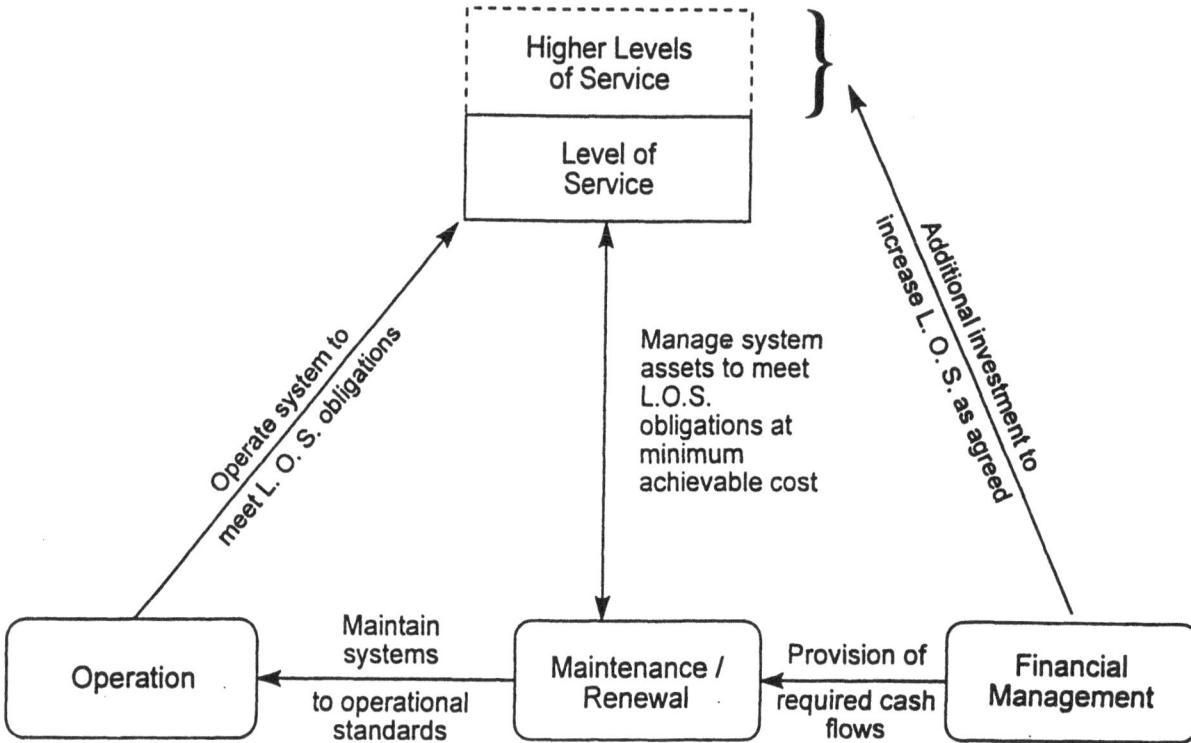

Note: For a specific Level of Service (L.O.S.), there will be an associated identifiable cost. In poorly managed systems, improvements in L.O.S. can be achieved by improving management processes and control (often by some additional costs and by re-ordering financial priorities). In well-managed systems, substantial increase in L.O.S. will generally require significant additional investment.

Figure 3 Financial Linkages for Financially Autonomous and Financially Dependent Agencies

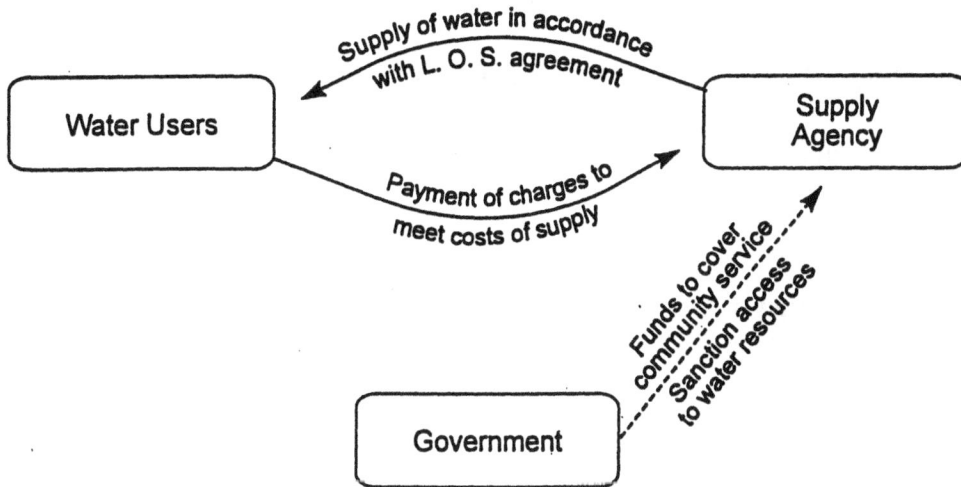

Water Users → Supply of water in accordance with L. O. S. agreement ← **Supply Agency**

Payment of charges to meet costs of supply

Government ⟶ Funds to cover community service / Sanction access to water resources

(a) Financially Autonomous Agencies
   (minimum government intervention)

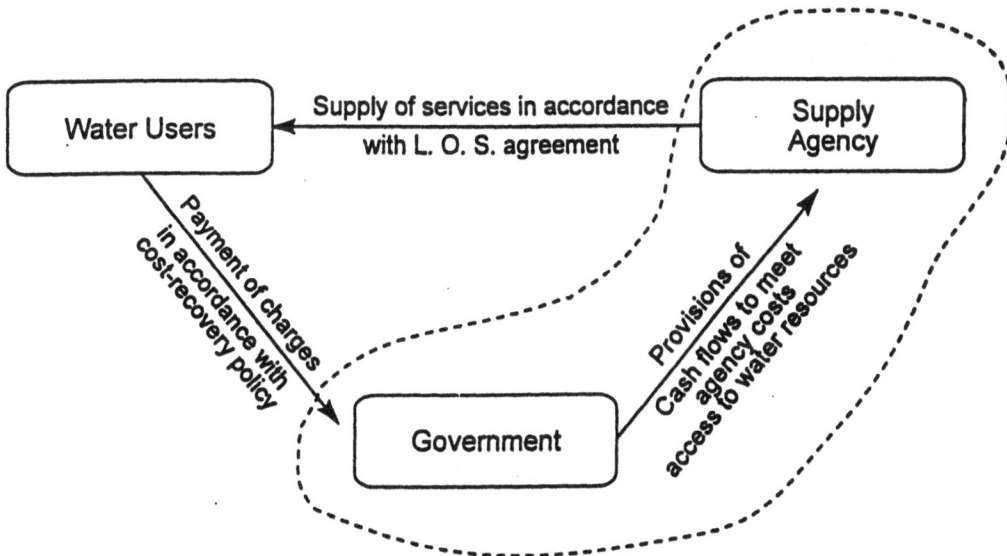

Water Users ← Supply of services in accordance with L. O. S. agreement ← **Supply Agency**

Payment of charges in accordance with cost-recovery policy

Provisions of Cash flows to meet agency costs / access to water resources

Government

(b) Financially Dependant Agencies
   (maximum government intervention)

Integration of POM with Institutional Planning Processes. The concepts embodied in an effective management process are universal and relatively simple, and can be described in the following four steps:

- identifying clear policies and objectives;

- developing and implementing programs and activities to achieve those objectives

- monitoring and evaluating progress, and identifying of gaps in performance;

- reviewing and adjusting objectives and programs.

However, each agency is in a unique situation, in size, client base, social and economic environment, managerial style and ethos, so the specific detailed management techniques need to be adapted accordingly.

In paragraphs b(i) and b(ii) above, it has been signified that agency, needs to develop the agency objectives which are in harmony with national planning policies.

For effective management, the government or other relevant authority must clearly state the organization's mission or purpose, powers, and functions for which it is to be accountable.

The senior management of the organization should develop by the process described below, a number of key objectives to be reached in carrying out the mission, and provide clear guidance to staff in the development of strategic programs and activities.

These policies and programs will be set out in a number of statements as plans, and companion documents to the POM.

These plans constitute the outcome of an integrated planning and management process. The process should provide for a "top-down" approach to priority/direction setting and a "bottom-up" approach to devising detailed strategies and activities. In this process, responsible managers within the organization prepare proposals for programs and activities to achieve the objectives and goals established by senior management.

Once approved, these proposals constitute the plans which the organization will implement. This integrated or corporate planning process, as it is known, provides for the various planning, operations and supporting activities across the organization to be brought together.

This is done by examining three basic questions:

- **Where are we now?**

An analysis for the organization of:

1. the current and forecast external environment in which it will be operating;

2. its existing strengths and weaknesses.

- **Where do we want to be?**

The formulation of:

1. a statement of purpose (mission and broad objectives)

2. a set of hierarchical objectives;

3. strategic objectives;

4. tactical objectives (operational objectives, i.e., specific, realistic, achievable, but challenging objectives with time-based targets)

- **How do we get there?**

1. Formulation of strategies, programs and activities directed towards achieving the established and approved objectives.

2. Objectives essentially reflect values or desired outcomes. In the context of managing irrigation distribution systems, the development of the objectives for the organization requires an understanding of the needs, demands and expectations of water users and landholders. The objectives are hierarchical in order, all relating to a view of the future, which will range from a broad view at the highest level to a detailed, specific outlook at the activity level.

3. The corporate planning process provides a systematic, integrated approach to the management of the total range of activities of an agency. It is an interactive process, involving for each specific objective:

    (a) identification of key result areas, i.e., activities which most significantly affect the level of overall performance in the area under review;

    (b) diagnostic analysis and identification of options for improvement in those areas;

    (c) development and implementation of programs;

    (d) monitoring and evaluating progress against agreed performance criteria;

(e)   periodic review of plans and programs.

References *(55)*, *(77)*, *(78)*, *(79)* and *(80)* provide additional detail of these processes.

The Corporate Plan. The full set of management plans comprises the overall corporate (or strategic) plan for the organization. A key requisite to the development of the effective corporate plans is the set of memoranda of understanding giving a clear assignment of functions, funding responsibilities, control and regulatory roles that may be divided between the irrigation organization and others, including government.

The corporate plan provides the basis and priorities for the coordinated development of annual work plans and budgets across the organization.

The corporate plan will normally set the directions for the organization for the short- to medium-term, say 3 to 5 years, and must be updated periodically, generally on an annual basis.

A typical set of management plans might comprise the following: *(refer to Figure 4)*.

- Development Plan

    1. new services to be provided;

    2. new facilities, programs or activities;

    3. major modernization or augmentation of existing facilities;

    4. services to be diminished or discontinued.

- Plan for Operations and Maintenance

    1. a permanently maintained set of documents and instructions, work procedures, programs and schedules;

    2. as described in paragraph B (a) above

- Research and Development

    1. operations research;

    2. materials research

    3. new or improved irrigation technology

Figure 4   Outcomes from Strategic Planning Process

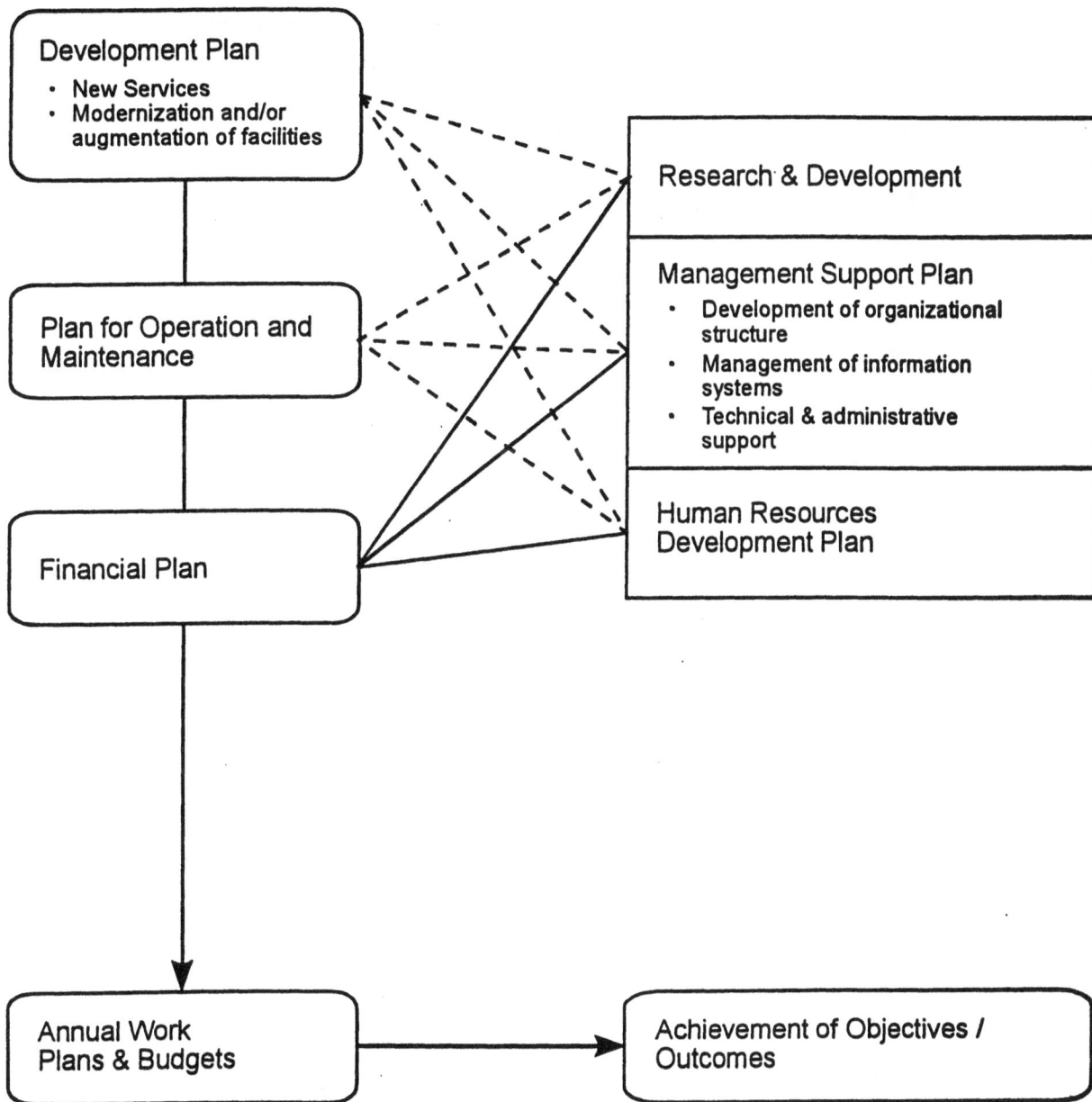

- Management Support Plan

  1. development and review of organizational structure;

  2. development of management information systems

  3. technical and administrative support;
     size, type and location of offices, depots and equipment to meet perceived needs.

- Human Resources Plan

  1. categories and levels of resources to carry out specific tasks;

  2. new skills required, or existing skills no longer required to address future activities;

  3. training needs analysis based on programs in the three plans above.

     ...regular or routine ongoing training programs, e.g., induction, skills training, management training;
     ...special programs to meet specific needs.

- Financial Plan

  1. expressing the extent of the organization's agreed programs in monetary terms;

  2. indicating sources of funds and cost recovery policies and targets;

  3. indicating strategies for dealing with shortfalls and emergencies, and generally for promoting financial efficiency.

For organizations or organizational units charged only with the responsibility for O&M of a particular water delivery system, the dominant plan will be the POM, and there may be no development plan, while the other elements of the corporate plan may be formulated by the parent institution.

Annual Work Plans and Budgets

The aggregated budget for the organization and the associated work plans reflect the agreed priorities of the organization. An annual work plan, should provide an annual update of the corporate plan and especially the POM.

FORMULATION OF POM

Using the Guide for Production of O&M Manuals

Managing the Process. The POM is described as one of the set of management plans, being a permanently maintained set of documents and instructions, organization charts, work procedures, programs and schedules, updated when changes are made, so that it comprises a complete statement for reference and guidance at every level in the project organization (*refer to Section B "PLANNING FRAMEWORK" of this document*).

A major component of the POM will be a manual comprising a number of sections dealing with each of the activities and functions. For larger projects, separate volumes for individual sections will prove more workable, and many sections will have several individual manuals.

For new projects, the production of O&M manuals should proceed through design and construction phases of the project implementation so that manuals will be available to operating and maintenance staff when the project is commissioned.

For existing projects where no comprehensive set of manuals currently exist, the production of the manuals will involve a number of individuals within the organization, sometimes also involving people with particular expertise from outside the organization. For all but the very simple systems with relatively few control structures, the production of manuals will extend over a considerable period, even years in some cases.

In either situation, it will be necessary to assign within the organization, the responsibility for managing the production process. This is distinct from the task of writing the technical contents of the individual sections referred to above.

The activities which need to be coordinated include:

- fixing the broad scope of the manual and its individual sections;

- establishing guidelines for format and style, so that individual sections can be clearly identified as being part of the set (providing "a corporate image");

- fixing a time schedule for completion of sections;

- nomination of authors of individual sections;

- controlling the printing and distribution of completed sections;

- arranging for review and updating.

Remember, these manuals form the essential building blocks on which the operational procedures, work programs and consequential annual budgets will be developed. They are vital to the successful formulation of an effective POM.

Steps in Developing the Manual

Step 1. (1) a clear mission statement or statement of the essential purpose of the organization is available;

(2) the type and nature of the organization is understood;

(3) the legal standing, and relationship and linkages to other units or organizations is clearly defined.

Step 2. With the information in Step 1 in mind, check the introductory parts of the guide, specifically, "CONTEXT AND SCOPE OF THIS GUIDE" and "PLANNING FRAMEWORK," to assist in interpreting and using the relevant information set out in Chapters 1 to 8.

Step 3. Assign responsibility for coordination as outlined in paragraph (a) (i) above "Managing the Process."

Step 4. For new projects, read Section B (b), paragraphs (i) to (v) and proceed accordingly.

Step5. For existing and new projects read Chapter 1, and set out the functions and management framework relevant to the particular organization or unit.

Step 6. Read Chapter 2 and establish the catalogue of facilities for which the organization or unit has responsibility for operation and maintenance.

Step 7. For the range of functions and category of facilities determined steps 5 and 6, apply the appropriate provisions of Chapters 3 to 8, in developing the relevant contents of sections of the manual, in accordance with the timetable and procedures established in step 3.

Procedures for New or Modernized Projects

The task of the O&M unit in formulating the POM must commence very early in the project planning phase, and will continue through the planning, design, construction, commissioning and operational phases.

Project Planning Phase. Project O&M must be addressed by knowledgeable O&M specialists in a comprehensive manner during project planning in the same way as any other project aspect such as selection of the conveyance system or determination of the agricultural activities.

In particular, the optimization of O&M/capital costs is vital, and the planning process should clearly expose the extent of any trade-offs in future O&M costs in planning options before final decisions are made.

Planners must recognize that every decision pertaining to the farm services or system facilities directly affects the long-term operation and maintenance function and its costs. Of primary concern in planning should be:

- the operational feasibility of the scheme relative to services intended and facilities selected;

- realistic costs of operation and maintenance to assure continued project integrity;

- the specific O&M facilities, communications, equipment, complement of parts and supplies;

- the advance staffing and training, and pre-transfer preparatory work of O&M that must be completed before the project commences operation. The substantial capital budget item for O&M must be carefully estimated.

The O&M matters that must be addressed during project planning and be fully reflected in the project feasibility report include:

- irrigation, drainage and flood control services to be provided to farmers and the related services to villages and Municipal and Industrial (M&I) customers;

- water allocations to individual farmers and customers and any interim modifications to utilize surplus water during project build-up;

- role of farmers in determining specifics of irrigation scheduling and system operation and maintenance;

- organizational structure of the O&M unit including geographical boundaries of the functional sub-units;

- data collection needs for purposes of O&M, extent of remote monitoring and control, and basic communication needs;

- configuration and citing of offices, shops, storage areas and housing;

- complement of fixed and mobile equipment, including back-up supplies and spare parts;

- schedule for completion of O&M facilities, procurement of equipment and supplies, and placing and training staff to meet scheduled start-up operations;

- cost estimates of O&M facilities, equipment and back-up;

- cost estimates of initial staffing;

- cost estimates of annual operations, including salaries, supplies, utilities, vehicles and allowance for staff replacement and training;

- cost estimates of annual replacement and maintenance of system facilities, equipment and buildings.

Design Phase. Aspects of project O&M that are addressed during the planning phase must be finalized during the design phase. These relate to:

- detailing the scheme of operation (i.e., controlled-volume, free draining, remote/on-site control, etc.);

- design of the overall conveyance/delivery system;

- the control, monitoring and communications system;

- the specific O&M offices, shops, yards and related features.

Preparation of procurement documents for the O&M equipment will have to be completed. At the same time, new tasks must be started. The specific O&M matters that must be addressed during the design phase include:

- procurement documents for the initial complement of O&M equipment, supplies and spare parts;

- detailed schedule for placing the system into O&M status and related actions;

- final cost estimates for annual O&M costs;

- staffing of initial project O&M personnel.

Construction Phase. Several aspects of project O&M must be pursued during the construction phase. In addition to those noted under the design phase, the dominant areas will be:

- installing the O&M organization in the field;

- commissioning of project facilities;

- transferring responsibility from construction to O&M.

Due to the typical staging of the project completion, construction will be underway in some areas, while full O&M will be in place elsewhere.

The specific matters that must be addressed include:

- finalizing and distributing the POM together with other documents;

- recruiting, placing and training O&M staff prior to start-up in accordance with schedules;

- farmer groups, if such are to be established, and elected/designated officers;

- orientation and procedural meetings with farmers and farmer groups;

- trial operations--internal and with farmers;

- trial maintenance--internal and with farmers.

Project Commissioning Procedures. Following commissioning of a new project, the O&M unit accepts full responsibility for the operation, maintenance and management of the completed project facilities. However, it is necessary that the O&M unit will have been involved in considerable preparatory work in the formulation of the POM.

Besides the preparatory work by the O&M unit, several documents are to be prepared by other units in the irrigation agencies.

The documents include:

- project feasibility plan

- designers criteria

- designers instructions to O&M

- right of way instructions for O&M

- construction/supply contract documents

- as-built drawings and manufacturers instructions

- facilities commissioning procedures

- initial complement of equipment and supplies

- initial complement of staff

Inputs to these documents will be required from the O&M perspective by the initial O&M unit staff assigned that responsibility.

Further details of these documents are set out in Annex 1.

Operational Phase. The implementation program, including timing, consists of a clear description of the activities required for the phasing-in of project O&M.

The issues to be included are:

- completion of system facilities

- commissioning of components

- transfer from construction to O&M

- commencement of services to each area

- preparatory O&M tasks, including:

  1. detailed work plan

  2. completion of O&M facilities

  3. equipment procurement

  4. staffing and training

  5. start-up procedures for services

- ongoing program

The matters to be resolved are discussed in Chapters 1 to 8 of the guide.

Update for Subsequent Project Stages. In some cases, a large project may be implemented in stages. Not infrequently, a considerable period may elapse between the commissioning of one stage, and the commencement of planning for the new stage.

The experience of actual operation of the initial stage or stages is invaluable in planning and implementing later stages.

All of the preliminary activities outlined above in paragraphs (i) to (v) should be completed in the sequence listed for the facilities involved in the new stage.

CHAPTER 1: ORGANIZATION, MANAGEMENT AND RESPONSIBILITIES

Effective management of an organization, or unit, requires a clear statement by the relevant management authority of the unit's mission or purpose, and those functions which the organization is to perform, and for which it is to be accountable.

The mission statement should provide a concise statement of essential purpose for which the organization or unit has been established. This statement should provide a clear indication to people within and outside the organization, as to the end to which the main thrust of organizational effort is directed. For example, the mission statement for an organization managing a discrete system to supply irrigation water to farms in the project area might read as follows:

"To operate and maintain the project facilities to supply crop water requirements to farms within the project area."

From time to time during the operational life a project, circumstances may arise which require a particular management effort directed towards achieving a specific outcome over a period of some years, such as:

- intensifying the cropping system;

- changing the cropping pattern;

- modernizing the system;

- controlling salinity;

- controlling the water table;

In such circumstances, this change of emphasis might be reflected in a review of the mission statement, as part of the process of refusing organizational effort to meet a newly perceived need.

The purpose of this chapter is to specify, for the unit responsible for operation and maintenance of the whole or part of a physical system, the management framework within which it will perform the functions assigned to it. Annex 2 contains some examples.

The following issues need to be addressed:

- pertinent project policies, within which the O&M function is to be exercised;

- functions;

- objectives and goals;

- functional units and responsibilities;

- detailed organizational structure;

- relationship with other public and private organizations;

- public relations.

PERTINENT PROJECT POLICIES UNDER WHICH THE O&M FUNCTION IS EXERCISED

The policies governing the access to water resources, the conveyance and distribution of water and the relationship of the organization with the farmers should be reflected here. Often these policies are condensed in a legal or contractual document which is the "Rules and Regulations" document discussed in Chapter 6. Where this is not the case, it will be important to include them here.

There may be two elements embodied in the provision of an irrigation supply:

- access or title to the water itself, expressed as a right to a specified volume on an annual basis--commonly called a "Water Right;"

- collection, conveyance and distribution of water to farm boundaries, in accordance with agreed rates of supply and delivery periods, either with or without limits to total volume of supply.

Depending on particular national policies and customs, and general availability of water, a formal process to allocate water rights process may not exist. Where such a process does exist, normally it has been adopted where water supplies are limited, and/or where there is competition for their use. Water rights may be granted by administrative action of a central water agency acting on behalf of government, or acquired under process specified by water law.

Basically the set of policies should cover the following main issues:

- The water rights or rules governing the access to water by individuals and the organization. Where the access to water is not regulated by water laws, indicate how the entitlement to water is defined;

- The main criteria that will govern the water allocation and distribution. Particular attention is to be paid to the measures to be adopted during emergency and droughts, and the priorities to be applied generally to the storage and supply of water under more normal circumstances;

- The rights of the organization and those of farmers to dispose of excess water;

- The rights of farmers to use excess water;

- The criteria to be used in the maintenance programs, such as the use of contractors, casual labor or machinery;

- The main criteria that will govern the relations with the water users;

- Regulatory provisions and disciplinary measures.

FUNCTIONS

Here the functions for which the O&M organization will have sole or major responsibility will be spelled out. Generally, the following functions are to be covered here:

- establishment of policies

- management, overall direction and coordination

- water resources, securing supply

- water distribution, including protection and security of source

- maintenance of facilities

- planning and design

- administration

- programming and budgetary control

- financing and auditing

- monitoring and evaluation

- safety.

Additional functions:

- flood control

- navigation

- recreation

- power generation

- fish and wildlife enhancement

- water supply for municipal and industrial use

- assistance to farmers on irrigation practices and on-farm development

- judgment and punishment of offenses made against the rules and regulations

- collection of fees and charges and other special functions.

The meaning and extent of those functions should be spelled out here. For instance, it may be necessary to specify that:

- the establishment of policies refers only to those that are related to the operation and maintenance of the systems;

- the planning and design refers only to the improvements and rehabilitation works that will be carried out within the context of maintenance programs;

- some of the above-mentioned functions are carried out by other entities or units outside the O&M organization, (as could be the case with the monitoring, evaluation, and auditing that are often carried out by external or independent units);

- certain functions will be carried out by the private sector or by contract.

How these functions are carried out, and by whom, is specific to every project. For instance in the private sector, the establishment of policies is the responsibility of a board of directors elected by the farmers. In public irrigation projects, the policies are sometimes dictated by higher levels of the organization or special committees often in consultation with water users organizations.

However, it will be important to specify which of the above functions will be carried out by the O&M organization, and those that farmers are expected to undertake by themselves.

Annex 2 has a number of examples of organizations and related functions statements.

GOALS AND OBJECTIVES

The goals and objectives, as here referred to, are those of the organization that deal specifically with the operation and maintenance of the physical systems (irrigation, drainage, roads and buildings). This organization may be one in itself or may be part of the overall project organization, in the latter case it will be important to describe the relationships and hierarchical dependencies between them.

The objectives should be described in as much detail as possible, classifying them in short- and long-term where possible. Typical long-term objectives of an O&M organization are:

- to provide a "satisfactory" operation and maintenance of the physical facilities of the project. "Satisfactory" is used here to designate the concept that regardless of the methods used to deliver water and to maintain the systems, the water users must find them acceptable;

- to maintain the system in "satisfactory" operational condition in perpetuity, conforming with the original design or approved design modifications. (the term "satisfactory" operation and maintenance is synonymous with agreed, approved or negotiated level of service);

- to provide that "satisfactory" service at minimum achievable cost, and depending on particular national policies;

- to recover costs of operation and maintenance from beneficiaries.

Depending on local circumstances and the actual range of functions, other objectives may be added relative to those additional functions, for example:

- allocating available water resources to different users within the project boundaries;

- controlling groundwater abstractions;

- establishing priorities for water use, etc.;

- collecting fees and water charges.

The short-term objectives should be described as specifically as possible, together with relevant time-frame for their achievement. These will refer to discrete activities directed towards the achievement of the longer term objectives.

FUNCTIONAL UNITS AND RESPONSIBILITIES

Every unit of the organization should have a clear description of its responsibilities. <u>This is an essential requirement for the proper functioning of the organization.</u> It is important to record not only the functional responsibilities but also the geographical coverage.

The number of units for which these responsibilities will have to be described depends on the complexity of the organization and the number of essential functions that will be carried out in each case. In most cases, a description of responsibilities will be required for the following:

(a)   general direction (board of directors, commissioners, special committees, director-general, general manager, chief engineer, etc.);

(b)   executive director, or project manager;

(c)   operations department. (this could include other units like water measurement, water distribution, etc.);

(d)   maintenance department and its subdivisions by hydraulic sectors or type of works to be executed;

(e)   administration and finance department;

(f)   farmers constituencies (general assembly, consultation bodies and others).

Several functions providing direct assistance to the project manager will be required to support the "line" functions outlined (a) to (f) above. These are frequently defined as "staff" functions. Frequently these are provided by individual staff members. Special training and in-depth experience are essential to assure the high level of advice sought by the manager. On smaller projects these functions may be provided by higher level state organizations or even private individuals. Regardless, a concise statement describing the function and all subordinate activities, reasons for providing the assistance, and any related procedures must be prepared. The staff functions would embrace:

- legal
- internal financial audit
- project performance evaluation
- safety
- environmental monitoring.

Where the organization undertakes additional functions, specific units may be required to deal with them, like:

- on-farm irrigation and drainage department

- laboratory service

- and others, as may be required.

DETAILED ORGANIZATIONAL STRUCTURE

The next step is to describe how the above functions and duties are discharged by organizational units and how the dependence and lines of authority are established. The most effective way of presenting this information is by an organizational chart with the necessary annotations.

A few observations appear relevant in this context. Experience all over the world is proving that too little attention is paid to the establishment of evaluation and monitoring units, but they are of great importance in assessing the performance of irrigation projects. A question that needs particular attention is whether operation and maintenance should be undertaken by a single unit or by two separate units. (*For an extended discussion of this issue see [Ref. 55].*)

Some examples of organization charts are included in Annex 2.

RELATIONSHIP WITH OTHER PUBLIC AND PRIVATE ORGANIZATIONS

The organization that manages the physical facilities of an irrigation project is sometimes part of a much larger organization that provides many services to the users, and in this case, it will be necessary to describe the institutional links with the larger organization. However, even when this is not the case, it is important to describe the links with other organizations, such as:

- land management
- research
- extension
- hydrological assessment
- credit schemes
- environmental and recreational agencies.

These links should be described, indicating the extent of the information, cooperation, or services provided, as well as the channels of communication between the respective organizations.

PUBLIC RELATIONS

Good public relations involve good communication. Here the communication channels between the project organization and the water users are to be detailed. Particular attention will be paid to the need for reaching all the water users and giving them the possibility of addressing the O&M organization when necessary.

The other aspect that also needs attention is the communication between the irrigation organization and the general public. This implies the use of mass media and other means to promote some efforts in the farming community or to pass relevant information in a rapid and effective way.

CHAPTER 2: PROJECT DESCRIPTION

GENERAL PROJECT FEATURES

The general project features and service areas should be described to facilitate understanding by all individuals involved with the project O&M.

A map or series of maps should be included to indicate:

- topographic features
- roads
- utility lines
- communities, and
- any other general project area features that will be important for O&M operations.

Specific project features will also require location maps. The irrigation and drainage system layout should be placed on these maps with details of:

- distribution points
- branches
- water measurement facilities
- crossing locations
- dams
- other water storage areas
- pump stations
- evaporation ponds
- maintenance shops
- offices, and
- any other pertinent details.

PROJECT FACILITIES

Detailed descriptions of all project facilities that will be operated and maintained by the organization will be necessary. The related project facilities that may affect O&M efforts should also be described to provide information that will facilitate efficiency and effectiveness of work. Details required include:

- specific location
- capacities
- operating ranges
- sizes
- unique features
- materials, and
- any other pertinent descriptions.

Some of these specific features are:

Water sources

- storage dams
- diversions
- wells
- facilities for mixing drain water for re-use.

Water distribution facilities

- canals
- pump stations
- pipelines
- siphons
- turnouts
- water level and flow control structures
- water measurement devices
- spillways, and
- related transmission and communication facilities

Drainage system

- outlet facilities

- pump stations

- main drains

- lateral drains

- bridges

- subsurface pipe drains

- culverts

- water table observation wells

- dicks, and

- water entry structures along drains.

Flood protection banks

Supporting infrastructure

- roads

- utility lines

- maintenance shops

- material storage areas

- offices

- equipment yards

- weather reporting system

- hydromet system

- parts depots, etc.

Other detailed design criteria, geologic reports, as-built drawings, etc. should be referenced in this section to be certain everyone knows where they are filed, and so that everyone can refer to them for specific details when necessary. The use of modern geographic information systems tools can greatly assist in this task.

CHAPTER 3: SYSTEM OPERATION

GENERAL

This chapter will provide specific, concise but detailed instructions for the operation of the irrigation system. It is to be used predominantly by operators in the field, and their supervisors and managers.

It will provide a formal documentation of operational procedures to assist in effective day-to-day operation, as well as providing a basis for longer term review and evaluation of policy and operational practices in the light of operational experience.

Two fundamental factors will influence the content of these instructions:

- the method of water allocation and distribution adopted for the system;

- the technology adopted for water control within the distribution system.

As emphasized in the introduction, section C, "Formulation of the POM," essential features of project operation should have been addressed during the planning, design and construction phases.

It follows, therefore, that the detailed instructions in this chapter should be compatible with the design features of the project facilities and the service standard specifications agreed/approved for the project.

There are a number of activities to be addressed in the formulation of system operation rules, which could be grouped under the following headings:

- detailed operational rules and specifications;

- irrigation plan (seasonal and annual operating plan);

- operational procedures

- emergency procedures

- operations below farm outlets.

DETAILED OPERATIONAL POLICY, RULES AND SPECIFICATIONS

Here the essential specific policy guidelines and general operating criteria which system operators must take into account in determining detailed operational procedures will be set out.

These will be extracted and expanded, where necessary for operational purposes, from the relevant information contained in Chapter 1, "Organization Management and Responsibilities," and Chapter 2, "Project Description."

They will include such matters as:

- water sources

    - any legal limits to water availability for project purposes;

    - any water sharing agreements with external bodies or organizations.

- priorities for delivery

    - normal availability from sources

    - restricted availability from sources.

- categories of demand to be supplied

    - project requirements

    - municipal and industrial (M&I)

    - environmental flows

    - recreation flows

- requirements for "passed-down-river" flows to meet riparian entitlements, or entitlements of downstream projects or water users, in terms of either flow rate or water levels to be maintained.

IRRIGATION PLAN - SEASONAL AND ANNUAL OPERATING PLAN

This section of an O&M manual should provide specific instructions for preparing the seasonal/annual irrigation plan. The objective of the exercise is to match the water demand with the supply as closely as possible. This exercise is generally complex and reiterative, and the use of computers may simplify the calculations. The complexity of the process varies from case to case depending on the scope for manipulating water supplies to meet the demand.

The preparation of the irrigation plan includes the following main steps:

- estimation of water supply (security of supply)
  ... wet season
  ... dry season

- estimation of water demand of the users (derived from cropping or demand pattern);

- application of appropriate water allocation criteria and procedures;

- matching supply and demand.

The water demand is essentially determined by the expected cropping pattern, or strict allocation procedures in water short situations. Depending on the country's social, economic and other conditions, farmers may have free choice of their crops and timing of cultivation activity or, in other cases, cropping patterns may be strictly imposed by the government. Preparation of the irrigation plan should be in accordance with the particular circumstances.

This chapter should clearly define the rules to be adopted in matching supply with the demand. In irrigation projects where the management has control over the cropping pattern, a good method is the use of an approval form for individual farmers. When the management has no authority over the crop selection, the rules for sharing water deficits should be well defined, for example:

- extending the interval between irrigation;

- decreasing the amount of water given to irrigation;

- allocation of water to preferential crops.

A variety of well-known formulae exist for the calculation of crop water requirements which take into account effective rainfall, temperature, crop growth coefficients, etc. A critical factor in the derivation of net irrigation requirements is the overall water use efficiency. The efficiency assumed is normally a target value and therefore normally overestimate the actual efficiency. It is important to monitor and evaluate the operations to assess the actual value of conveyance, distribution and on-farm efficiency as discussed in Chapter 8.

OPERATIONAL PROCEDURES

A specific set of written procedures and instructions will be required for each operating feature or item (or class) of plant, as indicated in the following sections:

Water Sources and Storages

The sources of water should have been determined during planning and documented. The quantity available from sources should be determined (forecast and holdover) on a periodic basis so that supplies can be estimated and plans can be formulated by the supplier and users.

Many irrigation systems utilize a reservoir, often a part of a multi-purpose scheme, to store water during periods of high river flow for subsequent use during periods of low flow. The dam which forms the reservoir is often a major structure and must be operated under specific rules and procedures. These rules are usually formulated during the planning, design, and operational phases.

Because of the critical nature of the dam and reservoir to the success of providing an adequate and reliable water supply, specific rules should be documented and implemented for each dam and reservoir, including provisions for periodic inspection.

Since the planning, design, operation and maintenance of large dams is a highly specialized activity, irrigation agencies responsible for such facilities should refer to directions and procedures developed by the International Commission on Large Dams (ICOLD) and their national committees, and the relevant specialized organizations within the country. Particular regard should be given to the requirements for instrumentation, monitoring and performance in the context of dam safety.

Distribution of Water

The operation of a water delivery network may vary considerably depending on a number of water management factors, including but not limited to the:

- climatic conditions, particularly the rainfall pattern;

- degree of regulation of the sources of water;

- quality of the water, particularly the silt content;

- size of project;

- number and type of farms;

- number and category of other users;

- type of conveyance and distribution facilities (open channels and/or buried pipes, etc.)

- method of water distribution; e.g. on demand, or pre-arranged demand, under a rigid rotational system, or under continuous flow.

The actual distribution of water includes two distinct steps:

- the preparation of the irrigation system scheduling (indenting, ordering) at an interval to be determined;

- the operation of the delivery system.

Procedures for these two activities should be clearly and carefully defined in the O&M manual since they are vital for the quality of services to the water users and will involve specific field staff.

System Scheduling, Indenting, Ordering

The preparation of a system scheduling depends, as indicated earlier, on the method of water distribution and on the type of facilities. The water order for an individual farm or group of cultivators or other users can be placed by each farmer or group, or decided unilaterally by the agency according to a pre-established scheduling. The preparation of the water delivery schedule can be simplified or even eliminated when part of the system is operated on demand or is equipped with advanced water control facilities, such as for downstream control or centralized remote control. Difficult areas in preparing a delivery schedule are the estimation of water propagation time, water use efficiencies and effect of rain interruptions. Knowledge gained from prior operational experience should be used in refining estimates.

Standard forms should be prepared to facilitate the preparation of the system scheduling, such as forms for:

- individual demand at lower level of canals;

- aggregating water demand for lower level to headworks incorporating efficiency values at different levels of the system. Instructions to deal with rapid variations of demand due to rainfall, prepared jointly with the users should also be included.

The use of modern computational tools can effectively assist in the task of calculating and scheduling water deliveries in irrigation canal networks.

Operation of the Canal System

Instructions should be formulated regarding:

- system start-up and close-down;

- range of discharges in each canal (minimum and maximum values);

- authorized rate of change of discharge;

- water level fluctuations at critical points of each canal (minimum, maximum, rate of fluctuation - normal and emergency);

- operating during rainfall season;

- operation of all water control structures (cross-regulators, offtakes, wasteways, pumps, etc.).

If part of the system is operated under remote control, detailed instructions for system scheduling and operating should be prepared.

Depending upon the type of water control technology, forms should be prepared for recording flow and water levels at critical points of the irrigation system. This information is important for:

- calculation of actual water delivered and used;

- determination of actual water use efficiencies;

- providing data for improvements in the system;

- volumetric water charges where applicable;

- longer term review and evaluation of policy and operational practices.

Given the enormous volume of information on canal operation and water delivery which needs to be recorded, stored, monitored and analyzed, the use of computer-based management information systems is proving advantageous, if not essential, in many countries. Such systems need to be developed carefully to ensure that all the information needs arising from the water distribution function for other units in the organization can be met without the need for separate data bases. Careful attention should be given in the development of the computer programs to include these other needs, as well as to providing for effective operational management. Refer also to Chapter 5.B "Management Information Systems."

EMERGENCY PROCEDURES

An emergency preparedness plan (referred to as a disaster plan in some countries) should be developed for all facilities for which failure or malfunction could cause:

- danger to human life;

- substantial property damage;

- loss of production;

- disruption of other community activities.

Essential complementary parts of an emergency plan are the:

- establishment of emergency depots with immediately available stockpiles of materials for rapid repairs; and

- schedules of mechanized plant and equipment which would be available from the agency, or from other agencies in relevant areas.

Dams and Major Structures

Given the nature of the hazards involved in structural failure or malfunction, and the specialized technology involved in these structures, reference should be made to ICOLD and other relevant organizations for instructions in preparing the emergency plan, including inundation maps.

Other Facilities

For other facilities, a number of situations need to be addressed, e.g.:

- excessive rainfall and flooding;

- blockages or malfunction of gates;

- breaches or overtopping of canal banks;

- breaches or overtopping of flood embankments;

- obstruction of drainage structures;

- chemical spills and pollution of waterways.

The plan should indicate:

- action to be taken to minimize damage or risk to structures;

- action to minimize danger to life or other property;

- internal reporting processes to be followed;

- external communication and notification processes;

- liaison requirements with relevant authorities:

- civilian protection or evacuation

- traffic control and diversion

- flood routing procedures

- water quality issues.

OPERATION BELOW FARM OUTLETS

When water is delivered at the farm gate, the operation below farm outlets is the responsibility of individual farmers. However, when farms are small as in many parts of Asia, it is common for the project deliver water at group level. In that case, the cooperation and active participation of farmers is essential for efficient use of water. These require organization, skill and discipline. The responsibility for organizing Water User Groups (WUG) should be clearly defined. These WUGs could be organized in a formal or informal way. The overall responsibility is to distribute water among the farmer members within the area, and sometimes also to maintain on-farm facilities. The organization and responsibility of each WUG and the rights and obligations of each member, should be clearly defined in a separate document.

The distribution of water by the WUG is dependent on the supply of water in the main canal, laterals and sub-lateral being operated by the irrigation agency. It is therefore necessary for the agency to take responsibility for, and an active interest in, activating the farmers within the WUG.

The maintenance work, where assigned to the WUG and consisting mainly of weed and silt removal in ditches and small repairs to structures, should be carried out under the guidance of the irrigation agency.

COMMUNICATIONS

An efficient system of communication is necessary to make possible the flow of information required for operation within the system and between the project and the users,

A full management information system is usually desirable, and this can be used by those responsible for different aspects of an irrigation scheme, such as the extent and rate of planting and harvesting, and occurrence of pests and diseases.

Clear instructions should be provided to operating staff on timing and nature of data to be exchanged.

CHAPTER 4: SYSTEM MAINTENANCE

GENERAL

All policies and procedures for, and assignment of, maintenance responsibilities relevant to all system elements and maintenance functions will be included in this chapter.

The most visible function of an irrigation agency is the conveyance and delivery of water to the fields. However, sustained success in this function depends not only on the effective planning and execution of water distribution operations, but on a well-planned and executed program of maintenance for all facilities, including drainage and flood control facilities. That program in turn depends on well-developed support procedures.

Effective procedures, for example, for the acquisition, handling and issuing of stores and spare parts for plant and equipment are vital to success. Effective planning for maintenance, on the other hand, must also recognize the inescapable lead time involved in stores acquisition, particularly if overseas purchasing is involved.

The procedures in this chapter, therefore, must be compatible with the general administration instructions defined in Chapter 5, and with the operational instructions defined in Chapter 3.

This chapter may contain a discussion of the approach to maintenance policy. In particular, it may include a discussion and directions on the following:

- degree to which preventative maintenance, as opposed to identification and resolution of problems on an ad hoc basis, is to be relied upon;

- appropriateness of deferring maintenance on facilities for which plans have been approved for modernization or rehabilitation;

- approach to modernization of works during performance of maintenance activities, including the degree to which it is the intent to continually modernize the system, and the criteria for such decisions;

- the relationship between system maintenance, modernization and rehabilitation.

All these aspects of maintenance policy are best determined within the context of a comprehensive asset management program for the irrigation and drainage infrastructure.

For these matters, this chapter includes general directions as well as delineation of responsibilities within, as well as external to, the operation and maintenance organization.

Development of Work Plans

Routine maintenance, which includes all work necessary to keep the irrigation system operating satisfactorily, should be documented and detailed in work plans. These plans should also include the work to be accomplished for all elements of the system. This work may:

- be performed on a periodic basis;

- be identified annually to be included in the following years' work plan;

Data to be used in developing maintenance work plans may originate from:

- reports from field personnel;

- inspection reports from engineers;

- performance assessment as outlined in Chapter 8, and other data sources.

Detailed instructions covering the formulation, completion, timing, and contents of work plans will be included in this chapter. Matters to be included are:

- contents and format of work plans;

- period for which plan is prepared (e.g. one year or longer);

- definition and extent of work;

- estimates of cost;

- timing of work, schedules of programs;

- method of execution, internal or external contract;

- assignment of responsibilities for execution of work;

- priorities assigned with regard to maintenance policies;

- maintenance of services during work programs;

- deadlines for provision of data;

- submission of work plans, approval process;

- notification and liaison, where work may affect activities of other authorities and individuals.

The assignment of responsibilities, as indicated above, should be reflected in job descriptions and assignment of responsibilities and delegations of authority as contained in Chapters 1 and 5. The work plans may cover a one-year period as well as longer periods. The planning periods to be covered in work plans should be stipulated and be consistent with the general budgeting and planning approach outlined in Chapter 7 and the institutional planning processes outlined in the planning framework.

Special Reserve Funds (Contingency Funds)

It may be appropriate to include in the budget, a special reserve fund to be accessed to repair or maintain the system in the event of unforeseen needs. This may include damage caused by major disasters, such as floods, earthquakes or structural failures. This chapter should state the criteria under which such a fund may be accessed and other general provisions for maintenance and administration of an adequate fund.

Maintenance of Record Plans

A general policy on the storage and maintenance of as-built plans, right-of-way plans and the updating of these plans as they are modified during maintenance activities should be included in this chapter. The policy should include procedures as well as assignment of responsibilities. The responsibilities for storage and updating of design engineers' instructions for operation and maintenance should be assigned. These instructions may contain a general strategy for inspection and maintenance of particular structures or facilities. If not included in the design engineers' instructions, the general strategy for maintenance of particular structures should be completed and updated by assigned maintenance personnel and included in section B of this chapter, along with design engineers' instructions.

SPECIFIC MAINTENANCE PROCEDURES

This section should contain details of strategies, policies, standards, procedures, record of management provisions, and other information specific to the maintenance of each system element or group of elements. The listing contained herein is for guidance purposes only, and is not exhaustive. Other categorization may be more appropriate for specific projects.

Given below is a summary of some of the more important maintenance aspects of these various features. To obtain additional information and details and for a broader and more thorough discussion, the reader should research other reference sources. In some cases, equipment maintenance manuals for a specific project, or recommendations and literature from the manufacturer of specific components, will provide a handy reference. Other references are referred to below, however these references are not necessarily a comprehensive list and other sources should be researched, especially those dealing with the site specific conditions of the country and project.

Dams and Reservoirs

Since the planning, design, operation and maintenance of dams and other large structures is a highly specialized activity, irrigation agencies responsible for such facilities should refer to procedures and directions developed by ICOLD and its National Committees, and specialist organizations within the country. The use of external review panels at intervals not exceeding 5 years is generally necessary to support "in-house" activity.

The following listing of the problems and hazards which must be addressed in the maintenance of storage reservoirs is included to assist in these discussions:

- sedimentation and siltation;

- water quality;

- bank erosion and slope instability;

- vegetation control;

- recreational hazards.

Work plans will include programs for:

- monitoring;

- catchment erosion control;

- control of pollution sources;

- bank protection;

- vegetation control.

Dams for irrigation projects are usually constructed of concrete, earth, rockfill or a combination of these materials. They should be inspected periodically for evidence of:

- cracking and settlement;

- instability;

- abnormal seepage;

- erosion;

- possible undermining of the downstream toe;

- foundation damage;

- concrete deterioration;

- other possible endangerment.

Work plans will include programs for:

- concrete refurbishment

- gate maintenance

- seepage control

- foundation grouting

- rip-rap replacement

- maintenance of control facilities

Sufficient monitoring should be included in the work plan to determine the extent, the cause, the rate of deterioration, and the short- and long-term effects of the problem. In addition, dam safety policies developed and approved by the organization should be included. The provisions of the policy should be applied to each structure systematically. A broader discussion and additional details are provided in References (3), (12) (20), (51), (55), (60), (68), (69), and (76).

Open Canals (Channels)

Canals are generally excavated in earth or soft rock and may be either lined or unlined. Items to be considered in their maintenance are:

- erosion of bed and banks

- damage to banks from human and animal activities

- settlement and sloughing

- silting

- vegetation

- seepage

- lining

- sealants

- under drainage.

Work plans will include programs of:

- monitoring

- canal straightening, realignment

- bank protection

- dredging, silt removal

- lining repair

- vegetation control (chemical or mechanical)

- seepage control.

References (3), (13), (20), (23), (28-29), (31-32), (34), (48), (55), (59-61), (63), (68), (73) and (76) provide thorough discussions and additional details.

Structures

Most structures associated with irrigation projects are utilized for the conveyance, regulation and control of water. They contain both structural and hydraulic features and are generally constructed of concrete, stone and brick masonry, timber, metal, rock, and rock gabions.

Structures associated with dams and reservoirs are:

- spillways

- weirs

- sluiceways

- tunnels

- riparian outlet works

- power outlets

- irrigation canal outlets and headworks

- fishways

- reservoir pumping stations.

Structures associated with open channels or drains may include:

- headgates

- check drops

- turnouts

- syphons

- flumes

- road crossings

- silt traps

- wasteways

- pumping stations

- cross drainage structures

- drain inlets

- water measurement structures.

Structures associated with pipe systems or buried pipe drains may include:

- inlets

- outlets

- silt and sand traps

- standpipes

- pressure relief/air inlet valves

- manholes

- crossings

- pumping stations.

Problems associated with these structures and their maintenance and repair requirements are generally similar. General maintenance due to age, natural attrition, and design or construction inadequacies are important. Additional details are provided in References (3), (20), (28), (31), (48), (55), (63), (68), (76).

Pipe Systems

In pipe distribution systems the maintenance of the conduits as well as the many appurtenances, such as gates, valves, metering devices, etc., must be considered,. The maintenance of some of these items, such as pumps, motors, electrical controls and automation, are discussed later. Problems include:

- damage to linings and coatings;

- corrosion;

- separation of pipe joints;

- build up of material in the pipe and appurtenances.

Work plans will include programs for:

- monitoring

- cathodic protection

- cleaning

- joint repair

- lining refurbishment.

Additional details are found in References (3), (4), (34), (35), (76).

Open Drains

Drains generally suffer rapid deterioration in condition affecting performance levels, and require comprehensive maintenance programs. Problems may include:

- erosion

- settlement

- sloughing

- siltation

- vegetation

- seepage.

Additional detail is provided in References (1), (3),(5), (20), (23), (28), (32), (60-61), (68), (73), (76).

Buried (Pipe) Drains

The major problems requiring maintenance include:

- physical blockages

- organic or biological blockages

- chemical or mineral sealing and outlet restrictions.

Work plans will include:

- monitoring

- cleaning

- root removal

- cleaning and repair of outlet grills.

Additional information on maintenance of subsurface drainage systems can be found in References (3), (24), (29), (56), (74), (76).

Flood Protection Embankments

Flood protection embankments are facilities not normally required to perform their function on a day-to-day basis. However, there is generally little opportunity to carry out routine maintenance during periods of flooding.

In these circumstances, maintenance programs should be implemented to ensure the facilities are fully maintained prior to flood periods. Items to be considered include:

- erosion and slumping of banks;

- rip-rap protection;

- damage to banks from human and animal traffic;

- vegetation and tree-growth on banks;

- bank cracking and seepage at structures;

- erosion and structures;

- control and cross drainage structures;

- access roadways;

- flood warning systems.

Roads

Roads located within an irrigation project and usually adjacent to a canal require maintenance to allow access to project features by operation and maintenance equipment and personnel. Types of roads associated with irrigation systems include:

- all-weather paved roads;

- unsurfaced or gravelled roads;

- berm roads along canals and drains.

Work plans will include:

- grading;

- gravelling and surface maintenance;

- slope protection;

- culvert and bridge maintenance.

See References (3), (55), and (68).

Pumping Stations and Electric Power Facilities

Pumps, motors, pumping stations, and electric power facilities are used extensively in irrigation and drainage projects in areas, such as:

- motorized operation and automation of structure flow control devices;

- sprinkler irrigation systems;

- computerized management facilities;

- pumping stations for pipeline distribution systems;

- pumped drainage;

- pumpwells;

- transformers, switchgear (often maintained by the power utility).

The equipment is generally specialized and comprehensive specific instructions on care and maintenance procedures are required, together with specialized training for maintenance personnel.

Specific care and maintenance procedures for each individual piece of equipment are usually described in the bulletins, manuals, and instructions furnished by the manufacturer. These should be included in the manual. An adequate supply of commonly used spare parts should always be kept on hand to ensure continuous operation.

For additional information see references as indicated:

- for pumping stations, References (3), (20), (26), (30), (45) and (72);

- for electric power facilities, References (3), (30) and (45).

Irrigation Wells

Maintenance of irrigation wells is primarily concerned with alleviating deposit build-up in and around the well screen and pump (encrustation and biofouling) and preventing or slowing the rate of corrosion and deterioration of the pump, screen, and well casing.

After installation, regular maintenance of the well is required to obtain satisfactory performance and extended life. Monitoring of the well discharge rate, specific draw-down and water quality is very important in detecting problems before they progress to a point where the well must be abandoned. See References (4), (20) and (43) for additional information on maintenance procedures and the control of corrosion and encrustation.

In addition to monitoring, work plans will include programs for:

- hydraulic flushing and redevelopment;

- chemical treatment;

- pump maintenance.

Cathodic Protection and Protective Coatings

The protection from corrosion of buried pipelines and appurtenances, and exposed metalwork generally is a highly specialized activity.

However, the direct financial losses and loss in operational effectiveness caused by shortened service lives of fixtures and equipment due to corrosion and cathodic attack are substantial. They are significant enough in most projects to warrant a specialist officer, or unit, to be assigned the responsibility for developing relevant maintenance and preventative programs for all of the project facilities and components likely to be affected.

These programs will be incorporated in the relevant work plans across the organization.

Communications and Sensing Equipment, Radio Links, Remote Monitors

The maintenance of equipment in these categories is also a highly specialized activity, generally requiring a specialist officer or unit to be established to have responsibility for developing maintenance programs across the organization.

Where the opportunity exists, it is generally advantageous to enter into period service contracts with specialist firms or suppliers.

Routine maintenance procedures to be carried out by internal personnel should be developed from the manufacturer's manuals, and incorporated in the work plans for the relevant units.

On-Farm Irrigation Systems

Most irrigation organizations do not have responsibilities for maintenance of on-farm systems. The following brief discussion is included for completeness.

Various types of on-farm irrigation have vastly different maintenance problems. Sprinkler (*References 37, 45 and 46*) and localized systems (drip/trickle) (*References 10, 22, 27 and 34*) offer special maintenance problems, and the listed references provide guidelines for their solution.

Sewage Effluent Irrigation Systems

The design of irrigation systems for disposal of sewage effluent requires special considerations. However, even properly designed systems may have additional or increased maintenance problems as compared to conventional systems. Depending on the degree of treatment to be employed the organic and inorganic chemicals in sewage can be quite corrosive and can induce biofouling. There is also the health hazard to consider particularly in terms of human contact with untreated or partially treated effluents. Untreated sewage effluent may contain suspended solids which can build up in pipelines, and clog valves, emitters, and sprinkler nozzles. Additional information on effluent irrigation can be obtained from Ref. 34.

Equipment, Buildings and Other Facilities

An organization may require a wide range of assets from computers to buildings which must be operated and maintained efficiently and effectively. Each type of asset will have different problems and unique elements in the maintenance work plan.

CHAPTER 5: ADMINISTRATION

GENERAL

Project management entails the application, direction and control of project resources to achieve the objectives of the entity. Obviously, the effectiveness and efficiency of project management depends on the quality of staff, adequacy of facilities, equipment and funds, the institutional arrangements and the timely flow of management information.

The primary institutional aspects are addressed in Chapter 2 with the facilities and other matters treated in Chapter 1. Aspects of administration, one of the functions noted in Chapter 2, is detailed further in this chapter. Programming and budget matters will be treated in Chapter 7.

The administration function provides the essential support to project management, and the dominant project functions--operations and maintenance. The following fundamental elements of administration will be addressed:

- management information systems;

- personnel functions and responsibilities;

- stores procurement and inventory control;

- financial procedures;

- administrative support procedures.

MANAGEMENT INFORMATION SYSTEMS

General

Effective management depends on the timely flow of management data and information to assist in controlling the day-to-day activities of an enterprise, and to provide a basis for longer term review and evaluation of the progress of the organization towards achievement of its objectives.

Given the nature of an irrigation system, the interactions involved, and the wide-ranging implications which flow from the irrigation activity, there is usually a great deal of data and information collected by an irrigation agency which is of direct relevance to other organizations and landholders. In some cases, the irrigation agency is best placed to collect other information which is essential for effective performance of other agencies, e.g. land management agencies, agriculture departments and research institutions, core government planning and budgetary departments.

A great deal of data and information needs to be collected, processed and presented in a form suitable for review and action by relevant managers and responsible officers throughout the organization, and to other relevant bodies.

Managing the flow of data and information is a critical activity for all organizations, and for optimum results, the processes need to be integrated across the organization. This is done by the design and maintenance of a Management Information System (MIS). Reference should be made to the Introduction, section B "Planning Framework," paragraph (b) "Institutional Planning and Management."

The responsibility for operating and maintaining the MIS should be clearly identified. It is usually designated with the administration function. However, the design of the system must involve those managers and supervisors within the organization who need the information and will use it in the exercise of their individual management responsibilities.

System Definition and Information Needs

The information needs for each organization are generally specific and must be clearly identified. However, the following listing is indicative of likely requirements:

- management reports;

- financial reports, expenditure against budget;

- costing reports;

- physical system status;

- water availability, storage status;

- canal and drain status;

- water deliveries and drain discharges;

- personnel status;

- maintenance program status;

- inventory control;

- plant availability;

- statistical reports;

- annual reports;

- other information reports.

For each of the various needs, procedures need to be established and promulgated covering:

- data to be collected, sources;

- frequency and method of compilation;

- content, timing and frequency of reports;

- distribution of information;

- staff participation and responsibilities.

Content, Timing and Frequency of Reports

The content, timing and frequency of reports will obviously differ according to the function concerned, e.g. water control information will be required on a continuous basis, sometimes in a "real-time" mode, whereas some financial reports will be on a periodic basis.

For management reports, these characteristics will vary, not only with function, but also with level of authority involved. For example, a board of directors, or a director-general, will only be interested in a weekly water deliveries report, while a manager of operations will need to see a daily or hourly water deliveries report.

Financial reports, involving manpower costing, should be compiled on the basis of a discrete pay-period, such as weekly or monthly, or multiples thereof.

For Management Reports, one useful criteria for the timing, content and frequency of reports is that these requirements should be determined according to the management level to which the report is directed and should allow sufficient time to take corrective action, if necessary.

If such response is not possible, then the reports are more correctly classified as information or statistical reports. This distinction will be particularly important for manually compiled and operated MIS, to avoid excessive reporting effort for little productive result.

Computer-Based MIS

With the increased availability of computer of greatly increased power and capacity, there is the opportunity to achieve significant improvements in the effectiveness of the MIS, and to achieve economies in staff time and costs.

Where communication facilities exist to provide low-cost data transmission links, it is possible to link a network of such computers throughout an organization, including regional and local offices. Such networking also offers the possibility of managed, integrated data bases, and greatly improved internal communication processes.

Many of the design principles for a computer-based MIS remain similar to those for a manually compiled system, however many procedures necessarily will be different. When converting from a manual system, care should be taken in system design to take advantages of all the possibilities offered by the new technology, and not simply computerize existing manual processes. Such conversion should be in line with a strategy developed under the planning processes outlined in the introduction to this guide in section B "Planning Framework," paragraph (b) "Institutional Planning and Management."

In maintaining the system, it will be necessary to develop procedures for:

- updating system description and facilities:

    - hardware
    - software
    - data acquisition and storage
    - reporting and communication links

- controlling development of new or modified systems;

- approval process for additional equipment;

- costing of services to internal client units.

While a computer-based MIS offers great flexibility and scope, the absence of computer facilities should not inhibit the development of an effective MIS based on manual techniques.

PERSONNEL FUNCTIONS AND RESPONSIBILITIES

Scope of Personnel Function

Personnel functions and the related responsibilities are to be described in detail. Parent agency, or government-wide procedures that dictate project organization actions should be noted and included in this chapter. Some of the functions to be described include:

1. establishment of personnel policies and procedures;

2. development and review of organizational structure;

3. setting staffing qualifications and levels of sub-units;

4.   preparing position descriptions;

5.   preparing candidate requirements;

6.   recruitment/evaluation;

7.   determining compensation and benefit levels;

8.   staff processing/orientation;

9.   staff development and succession planning;

10.   maintaining personnel files;

11.   payroll;

12.   staff termination.

Responsibilities for initiating, directing and executing each function will vary. For example, the project manager will likely have final review of the first seven, and personally direct items 1, 2 and 3. Other functional managers--operations and maintenance--will likely participate in items 1 through 7, and item 9, and have on-going responsibilities for 4, 5 and 6. The project administrative manager will participate in all, helping to execute the first seven items and direct the remaining ones.

Personnel Policies

Staff recruitment, training, compensation, benefits, advancement, specialization, and field expenses are a few of the personnel policies to be described. These are to be written in terms for ready use by managers, staff, and the administrative unit. Policy formulation should be directed towards the objective of attracting capable people, providing for constant improvement and maintaining a highly skilled dedicated staff. Equally, policies should facilitate identification of unresponsive, low producing and otherwise ineffective staff, and identify processes for performance improvement or removal from office.

Personnel Procedures

Procedures necessary for executing the functions described under (a) in conformance with policies described under (b) must be set forth in a clear manner for guiding responsible staff in the executing of these activities. Procedures are to be set forth in terms of function area, objective, actions and the initiating, executing, participatory, review, and approval responsibilities as appropriate.

STORES PROCUREMENT AND INVENTORY CONTROL

Procurement Procedures

Typically, responsibility for acquisition of essential items needed by the project organization is assigned to one sub-unit. It in turn may submit requests to a centralized agency or government purchasing unit, or it may procure them directly from outside sources. The outside sources may be in the private sector or from government purchasing unit, or it may be in the private sector or from government or quasi- government entities.

Clear procedures are to be formulated covering every action of the acquisition unit from receiving requests, through processing requests, sourcing items, business transaction, receiving items and transferring/releasing them to the end user. Initiating, approval, executing and review/audit responsibilities are to be set forth. These procedures should ensure the disclosure and elimination of opportunities for actual or potential misappropriation of materials and equipment.

The classes of items to be acquired would include:

- equipment and spare parts

- materials

- supplies

- property

- support services--professional, maintenance, construction.

The timely provision of stores and services is of fundamental significance in achieving the objectives of the operation and maintenance functions. The adopted procedures for procurement and supply will be directed to that end, with automatic checks to monitor progress.

On the other hand, the end user needs to recognize the time involved in the supply process, which will vary for the different classes of item and the need for tenders, etc. (*see Annex 3*). The officers responsible for developing the POM and maintenance work plans will need to take these into account when planning their programs. This aspect is particularly important when overseas purchasing is involved. See also Chapter 7, paragraph F "Foreign Exchange."

A typical flow chart for the procurement process is included as Annex 3: Sheet 1 indicates typical supply period objectives, Sheet 2 indicates a typical process for an individual item.

Custody Issue and Disposal

Procedures for the receipt, care and custody of stores and materials should be formulated. The matters to be dealt with include:

- inspection and receipt

- payment and accounting

- stores facilities

    - security
    - central and local

- stores issues, imprest stores

- standard items--maximum and minimum holdings

- custodial responsibility.

Procedures for disposal of unused or redundant items are also to be formulated. This, should treat the same general classes of items listed in (a) above, together with procedures for the consequential financial adjustment, credits, write-offs, etc.

Responsibility for Project Services

It should be clarified whether or not project services, i.e., water supply, drainage, electricity supply and waste handling, would be handled by this unit.

Water contracts, water contract management, and the related financial matters would be under the responsibility of a financial sub-unit with direct participation and support of the project operations unit.

FINANCIAL PROCEDURES

Most financial procedures are standardized in government or by the central management of large or parent institutions. The associated rules, regulations, and accounting standards will be followed. But, nevertheless, some tailoring to the project may be needed as well as supporting procedures where standard procedures do not suffice. Procedures for authorizing staff who will have initiating and approval responsibilities must be defined. A complete set of procedures must be prepared covering all financial actions. These procedures should make auditing efficient and ensure probity in managing the financial affairs of the organization. These include:

- budget documentation

- budget processing

- drafts on funds

- invoicing

- receipts

- revenue collection (if applicable)

- deposits

- accounting

- payroll

- personnel financial records.

ADMINISTRATIVE SUPPORT PROCEDURES

Specific areas of support and the associated procedures are to be described in a manner that the line unit managers and staff will clearly know of the support and means available to secure it. These will also guide administrative staff in the execution of their responsibilities and related tasks. The actions required by all participants are to be described. Usual areas of support to be addressed include:

- travel

- office communications

- computer services

- meeting/conference facilities

- typing and clerical services

- records, correspondence files

- plan filing

- office equipment repairs

- office maintenance

- office supplies

- printing, reproduction

- media liaison

CHAPTER 6: WATER USERS

RELATIONSHIP BETWEEN THE PROJECT AND WATER USERS

This chapter of an O&M manual should deal with the relationships between an irrigation agency and the water users. It should clarify the rights and obligations of each party, which will depend on the adopted organizations, and the nature of the supply arrangements for the system, for example:

- system controlled by farmers;

- system controlled by government officials;

- system with parts controlled by government officials, parts controlled by farmers.

Whatever the arrangements for system management, it is vital that the relationships between the irrigation agency and the users be clearly defined and understood to provide the best service to the users. Preferably they should be condensed in a small document or brochure and distributed to all users. The cooperation of users is essential to the successful operation and management of an irrigation project. While the irrigation agency has the responsibility for distribution of the water to individuals (or, in many cases, to groups of water users) and for the maintenance of the conveyance and distribution system, the farmers are responsible for the operation and maintenance of their own farm facilities, and, in some cases, of the system delivery from the project delivery point to the individual farms. Mutual understanding and cooperation is essential for effective overall management.

RIGHTS AND OBLIGATIONS OF WATER USERS

The rights of water users derive from the project policies discussed in paragraph A of Chapter 2. However, the actual rights and obligations of the users should be presented in this section of the O&M manual explicitly depending on the type of organization. User rights may also include:

- participation in the election process of the representatives of water users;

- access to other services provided by the agency.

The obligations of water users may include but are not limited to:

- implementation of the approved cropping pattern;

- timely order of water and compliance with the scheduling for water delivery established by the agency;

- making best use of water on the farm with minimum of losses and without harm to other users;

- maintenance in "satisfactory" conditions of that part of the delivery system for which they have responsibility;

- cooperation with the agency in the works which are carried out for their benefit (maintenance or improvement works);

- payment in due time of any land/water or other charges and levies;

- compliance with effluent water quality standards or criteria, limiting the use of toxic materials, etc.

OFFENSES AND PENALTIES

As discussed above, it is important that water users, and project employees as well, comply with their respective obligations. The most frequent offenses by the water users and the resulting penalties should be spelled out and known by the water users to reduce their occurrence. They may include:

- abstraction of water without project authorization;

- non-compliance with the approved irrigation schedule at farmers field level;

- non-execution of maintenance works which are under their responsibility;

- non-payment of water charges;

- acts of vandalism, damages to project facilities and harm to other users.

OTHER SERVICES

Besides the delivery of water, project farmers may benefit from other services from the project agency which may include:

- technical assistance for on-farm water management and other activities;

- delivery of inputs (fertilizer, pesticides, seeds, etc.);

- Input to farming activities (plowing, fertilizer treatments, etc.);

- financial assistance.

The rights and obligations of the farmers regarding those other services should be presented in a separate document since it is not an integral part of an O&M manual.

CHAPTER 7: BUDGET DEVELOPMENT AND PROGRAMMING

GENERAL

The development of the budget is an important element in the planning and management process for any organization. The budget documents provide a forecast and express the commitment in financial terms, to the programs, works and activities that the organization intends to carry out in the period under consideration, usually one financial year.

The approval of the budget will be given at the appropriate level of authority or institutional level, and in most cases, approval at government level is also required, for all or part of the budget provisions. Accordingly, the procedures discussed in this chapter may include necessary interaction at government level. It is recognized that there are many privately managed irrigation projects where there is no formal government involvement in the O&M of the distribution system. However, most of the principles outlined in this chapter are applicable to both private and government projects.

A clear and concise presentation of the budget contents will assist consideration by the relevant authorities, and enhance the chances of a positive outcome for the organization. In this connection, it is absolutely imperative that the budget request in any year is framed to meet the agreed objectives for the organization and in accordance with its policies and priorities. Moreover, many of the budget proposals will have financial implications which extend beyond one financial year, and individual budgets need to be framed in the context of the organization's longer term financial plan. Refer to the introduction, section B "Planning Framework," paragraph (b) "Institutional Planning and Management," and paragraph (c) "Annual Works Plans and Budgets."

Once approved the budget provides the authorization and financial framework for work programs for the period.

Coordination of the Budget Development

The development of the budget will involve every unit in the organization. The coordination of the budget process is the ultimate responsibility of the nominated officer (budget coordinator), usually located within the administration/financial unit of the organization.

It will be this person's responsibility to develop and promulgate the procedures within the organization. These will be outlined further in this chapter. Where appropriate, these procedures must conform to those in context and timing set out by the central government agency, e.g. treasury and department of finance.

The Budgetary Cycle

There are a number of stages and discrete activities involved in the formulation, approval and implementation of a budget which are followed on a year-to-year basis. Collectively, they are generally referred to as the budget cycle. The following listing is indicative of typical activities:

Formulation of a budget request

- review and evaluation of previous budget performance;

- determination of key issues and priorities--consultation with water users;

- issue of budget guidelines to units;

- development of budget requests by units--consultation with users;

- aggregation of budget requests at organization level.

Approval process

- review and adjustment at:

- management level

- board level

- government level

- comparison of spending proposals with estimates of revenue

- approval of budgets.

Implementation of approved budget

- notification of approved budgets to units;

- adjustment (if necessary) of unit work plans and programs;

- implementation of approved plans and programs;

- monitor and review, adjust plans and programs as necessary;

- completion of programs, finalization of expenditure for period.

BUDGET PROCEDURES

Format and Timing

The budget cycle extends over three budget periods. The formulation and approval processes must be completed prior to the financial year to which they apply, and the final review and evaluation of budget performance can only be completed at the end of that year of implementation.

The formulation and approval period has fixed deadlines, often dictated by the government budgetary and appropriation processes. Accordingly, the units in the organization will need to have a disciplined approach to this activity.

The responsible budget officer will set out the timetable for the budget formulation, specifying the dates by which the various stages will be completed. Annex 4 indicates a typical timetable for budget formulation. Sheet 1 shows a corporate planning program.

As the individual unit budget requests need to be aggregated and reviewed at organizational level, there will be a need for standard forms and documents to be developed, both to facilitate their original formulation within the individual units and the subsequent aggregation by the budget coordinator.

The budget coordinator will prepare the budget forms, and related instructions and specifications in consultation with the appropriate functional managers, and arrange for their distribution.

The specifications and instructions will cover:

- description of activity or program;

- justification for activity:

> ... relationship to objectives for particular functional responsibility
> ... commitment by organization or government
> ... effect on the service being provided
> ... economic justification (if appropriate)
> ... priority
> ... whether a continuing or a new activity.

Scheduling of associated resource commitments will be expressed ultimately in financial cost. Typical items are:

- personnel:

    ...   numbers and classification

    ...   wages and salary costs

    ...   related allowances and expenses

- equipment and plant hire

- supplies and materials

- general energy, fuel costs

- pumping costs

- technical service costs--internal, external

- contract services

- training

- travel: domestic and international

- administrative and general expenses

- any other item of projected expenditure.

The instructions will also provide advice on standardized costs to be used in estimating:

- salary and wage rates

- fuel, energy unit costs

- plant hire, hourly or daily rate

- particular materials and supplies

- inflationary factors (if applicable).

The estimates of costs and resource commitments will be drawn from the estimates set out in the various elements of the POM and associated work plans, as described in earlier chapters.

If the estimate applies to systems or parts of systems which are not fully operational, then the estimates will be made for the planning, design and construction phases as outlined in the introduction, section C "Formulation of POM."

In some instances, an organization may also be involved in programs for new work or modernization, or additional functions funded separately from the O&M budget. In such cases, the budget documentation will provide for these estimates to be made separately from the O&M estimates.

Implementation of the Approved Budget--Budgetary Control

Once the budget has been approved the following actions will occur:

- the budget coordinator will advise organizational units of the approved budgetary allocations in each case;

- unit managers will adjust (if necessary) their work plans and programs in accordance with approved allocation of funds;

- unit managers will forward revised estimates to budget coordinator;

- unit managers will provide information on the progress of their work plans and programs, and periodically provide revised forecasts of their expenditure pattern.

- budget coordinator will monitor and provide periodic report on implementation.

The reports and reviews by the budget coordinator will conform to the specifications set out for the Management Information System (MIS). Refer to Chapter 5 "Administration," paragraph B "Management Information Systems" for details.

FUNDING SOURCES--COST RECOVERY

Measures for Cost Recovery

The national policies for funding O&M budgets will vary according to particular circumstances and government economic and social objectives. In view of the long-term nature of investments in irrigation infrastructure, questions of inter-generation equity frequently arise.

The range of measures adopted (sometimes in combination) by which revenue is raised, or provided to an irrigation agency to meet its costs, include the following:

- appropriations from government taxation or other revenues;

- land/water lease or rental charges;

- crop levy on farm production;

- water charges on water users, related to all or part of costs of operation and maintenance of distribution facilities;

- water charges on users, based on full costs of project facilities, including depreciation and interest on loans;

- water pricing to achieve a stipulated economic rate of return based on the life cycle costing of system assets or the value of the water to other uses.

Irrespective of the particular measure or range of measures adopted, the cost of O&M and revenue to meet its cost will be brought into relative context for either pricing or policy considerations. This will be done at one or more of the following levels:

- government level

- institution or agency level

- project level.

Funding Sources

Where the responsibility exists for cost-recovery programs at agency or project level, all revenue estimates and sources must be identified in the budget documentation.

Funding sources will include:

- federal entities

- state entities

- provincial entities

- local entities

- landholders/lease holders

- water users

- other users or beneficiaries.

FUNDING RELIABILITY

The costs associated with operating and maintaining an irrigation system contain fixed and variable costs. The nature of the enterprise is such that both fixed and variable costs are climate dependent, and frequently costs are incurred over which management has no control, and cannot be forecast reliably, e.g. droughts, floods. The need for maintenance effort can vary over the life of a project, depending on the age of the components relative to the anticipated service life.

On the other hand, it is possible for anticipated revenue to vary from "normal." If revenue is based on crop production, then periods of low revenue may occur due to poor yields or low prices. If revenue is based on the sale of the water, then the occurrence of droughts, or higher that normal rainfall reducing demand, will result in lower that normal revenue, even though actual operating costs may be higher because of these abnormal circumstances.

The budget process normally requires forward estimates to be based on "normal" conditions, unless it is known otherwise with some certainty.

The requirements for O&M functions do not have the same order of possible fluctuations as revenue sources. It is not possible to drastically curtail these costs in a year, or a series of years, without affecting the performance of the system. Lower performance means higher costs and/or reduced capacity to earn income now and in the future.

Therefore, some underwriting of cash flows to an agency may be required to equalize revenue generation. This can be done by one or more of the following methods:

- advances from government

- establishment of an equalization reserve fund

- overdraft facilities

It would be expected that the operation of these funds would be equalized by returns in better than average seasons.

SPECIAL FUNDS

There may be circumstances for the establishment of special funds, or special arrangements to access additional funds in particular situations:

Emergency or Contingency Funds

To provide for unexpected operating or maintenance costs during the course of a financial year, e.g. floods, failure of facilities.

Replacement Funds

If future significant expenditures are anticipated to be required to replace project facilities, it may be appropriate to include depreciation components in current charges or revenue, and place these amounts in a "replacement" account subject to established procedures.

Construction, Rehabilitation and Modernization

If programs of this type exist, they may be funded by external funding, which could carry some interest or redemption charge. If so, these charges should be included in the budget estimates, together with reference to the fund sources which are to meet them.

FOREIGN EXCHANGE

In some countries, problems of foreign exchange arise in funding O&M. Foreign exchange may be required for the purchase of equipment, materials, spare parts, for training or other activities.

In some cases, the O&M organization may purchase the necessary foreign exchange directly. Frequently, applications need to be made through the relevant government authority.

In any case, it is important to identify foreign exchange requirements well in advance of the actual procurement so that stores and other items can be replenished without creating shortages that adversely affect the timing and effectiveness of O&M activities.

WATER SERVICE CHARGES

The practice of prescribing a fee or royalty for the use of water varies for individual countries. In some, no fee is charged for water extracted by an individual from wells or rivers for human or stock uses, and in some cases such free access is being enshrined in legislation. In others, a fee is collected for access to a nominated volume of water, e.g. for water rights.

In many countries, moves are now being made to differentiate between a fee or royalty as described above, and water charges. Water charges or water rates are raised on consumers or landholders to cover all or part of the costs of a water supply system, i.e. those costs associated with the collection, storage and/or extraction of water from a source, and its distribution to consumers through a network of streams, canals, pipelines or aqueducts. Where all or part of the agency costs are met by water service charges, procedures will be established for raising the charge, its assessment and collection, and accounting procedures.

Estimates of projected revenue raising should be included in the budget documents.

Some of the factors involved are:

- legal authority for charges or fees;

- class use of water e.g. gravity supply, relift, pumped diversion;

- basis of charge --area, volume, crop type;

- basis of assessment/measurement, estimation, formulae;

- additional charges--excess water use, special timing of delivery;

- drainage charge;

- basic service charge.

CHAPTER 8: MONITORING AND EVALUATION

The monitoring and evaluation details required for projects vary considerably between irrigation and drainage projects. These details are most important to smooth, long-term, efficient system management and are critical in setting priorities for O&M and adjusting seasonal and yearly operational requirements. It is especially important to assessing the level of service that has been achieved and adjusting the O&M effort to meet the system management objectives as described in the Introduction B (b) and illustrated in Figure 2.

These activities are most effective if carried out at two levels. Normally the on-going detailed monitoring and evaluation is assigned to one or more units. This unit (or units) is responsible for processing and distribution of information from evaluations to appropriate managers that have need for the details. Another small unit typically carries out audits, both physical and financial, directly for the project manager's office to monitor and evaluate organizational performance. This chapter will deal primarily with the first activity.

MONITORING

Items to be monitored must be specifically noted. The organizational unit responsible for each monitoring activity must be identified. A monitoring plan include:

- Collecting data on the following factors:

    ... precipitation and temperature
    ... crop production (area, yields, type of crops)
    ... water quality
    ... water use--farmers, municipal, industrial, and others
    ... groundwater quality and levels
    ... return flows
    ... drainage water quantity and quality
    ... soil salinity

- operating costs of major components, such as:

    ... individual pumping plants
    ... main water supply
    ... distribution to blocks

- maintenance activities, schedule and costs for major components;

- monitoring locations for each activity;

- methods and procedures for each monitoring activity;

- timing of monitoring actions;

- data presentation, format, detail and storage;

- distribution of information.

EVALUATIONS

Evaluations of information gathered in the monitoring process must also be systematically performed. The organization units responsible for evaluating data must be identified and assigned specific evaluation areas. General information which is needed for each type of evaluation will include:

- data sources (from monitoring and other areas);

- timing of evaluations;

- who will receive the reports and when

- methods to be used in making evaluations for each purpose;

- format for the evaluations to be distributed, including reports to be prepared reflecting the purpose of the evaluation;

PERFORMING EVALUATIONS

To manage a system properly, the physical effectiveness of past operations must be considered against the original criteria set for the project, or as subsequently amended following modification of the facilities. This is often embodied in a set of levels of service. Procedures for acting on the matters uncovered in evaluations are critical to the financial and operational efficiencies of a system. Priorities for adjustments in the system and scheduling the needed maintenance can best be made using inputs from timely and proper evaluation reports. Some of the diagnostic analyses that can be considered are:

- farmers operational performance:

    ... adequacy of crop production techniques for irrigated farming;
    ... adequacy of irrigation methods;

- ... farm management and economic results;
- ... soil management and erosion control;
- ... on-farm efficiency of water use.

- delivery operational performance

  - ... water use efficiency for a distribution;
  - ... water losses (physical)-- deep percolation, canal seepage
  - ... spillage from canals
  - ... reservoir seepage
  - ... water operational losses
  - ... adequacy of delivery scheduling
  - ... energy use

- drainage operational performance:

  - ... drainage requirement change by area;
  - ... water table fluctuation by season and years;
  - ... water quality changes by area for drain effluents;
  - ... soil salinity changes by area;
  - ... occurrence and extension of flooding

- maintenance of individual components:

  - ... civil works
    - ° canals
    - ° structures
    - ° drains

  - ... equipment degradation and prediction of replacement schedule:
    - ° fixed (pumps, hoists, etc.)
    - ° moveable (earthmovers, transport vehicles, truck, loaders, etc.)
    - ° computers and office equipment

- overall project review: efficiency and effectiveness. The procedures outlined in the four main categorized above will evaluate the relative performance of various project components and activities, and should expose whether any poor performance is strictly a technical or managerial problem which may be resolved by internal management processes.

It may be necessary, from time to time, to carry out a more wide-ranging evaluation of the total project, for example, if poor performance is a result of inadequate flow of funds for O&M because of inadequate generation of benefits, or from external economic, social or environmental effects.

Some of the matters which should be canvassed in such a review are:

- documentation of project costs and revenues;

- adequacy of revenue sources to meet O&M needs;

- benefit flows from project to farmers, governments, others;

- comparison of benefits generated to revenue required;

- relevant agricultural and engineering issues;

- social and environmental changes and resulting implications;

- institutional effectiveness in providing efficient and effective system operation and services to water users.

ANNEX 1

GUIDE TO AUXILIARY DOCUMENTS:
PROJECT OPERATION AND MAINTENANCE

GENERAL

The Plan of Operation and Maintenance (POM) constitutes the comprehensive guide, detailed instruction, background information and documentation for operation and maintenance of a project. A description of the POM is provided in the introductory sections of the document, "Guide for Preparation of a Plan for Operation and Maintenance."

Besides the preparatory work by the O&M unit, several documents are to be prepared by other units in the irrigation agencies prior to commencing operation. These are to convey instructions and/or information to be incorporated into the POM with the complete documents serving as reference on the subject items. The documents include:

Project Feasibility Plan. The project planning document forms an important part of the O&M reference material. Of particular importance, in addition to the report, are the policies, rules, regulations and legislation bearing on O&M. Water rights and allocations are examples. And, of course, the details of the adopted project services, farmer obligations, cost allocation, water charges, agency/farmer O&M responsibilities and all other project commitments are essential to this POM.

Designers' Criteria. The design unit is to prepare a comprehensive report stating criteria used in design of the facilities. These are to cover such matters as material characteristics, allowable stresses, allowable loadings, allowable loading conditions on and adjacent to structures, protective measures to be maintained effective, and surface drainage removal.

Designers' Instructions to O&M. The design unit is to prepare a comprehensive report clarifying the permissible operating conditions including start-up and shut-down of each individual facility, system sub-component and system. Permissible rates of filling and emptying specific canal reaches, siphons and pipelines are obvious examples. Rate of operation of gates and valves are another. Required compaction and shape of canal prisms to be maintained, including cross-slope of road subgrade and road surfacing; inspection and performance of tow drains and bridge supports; and cautions when finding dampness at aqueduct abutments, concerns with surface drains and canal lining are yet other examples. This is an important document that the people who design must complete at the time plans and specifications for construction are readied, since others cannot reconstitute these guides nor are O&M staff capable or responsible for developing them.

Right-of-Way Instructions for O&M. Maps of right-of-way and conditions for relocation of utilities are to be documented. Specific provision of access or other factors affecting O&M are to be stated.

Construction/Supply Contract Documents. Sets of drawings and specifications for all works should be furnished to the O&M unit at the time of tender. Subsequently, copies of change orders shall be sent when issued. These will form the initial basis for preparation of O&M manuals to be further refined using as-built drawings.

As-built Drawings and Manufacturers Instructions. As-built drawings should be completed by the design unit and forwarded to O&M within six months of acceptance of a project component from the contractor. These are essential for completing O&M manuals and procedures and their receipt should not be delayed until contract completion for large contracts nor until all components are finished. Likewise, manufacturers/suppliers warranties and instructions on equipment and materials should be provided as received.

Facilities Commissioning Procedures. Specific procedures for commissioning individual facilities, system sub-components and systems are to be stated. These are to include a description of acceptance tests, start-up procedures, measurements and remedy of deficiencies. The participation of representatives from the design, construction and O&M units are to be identified with responsibilities clearly stated. The construction unit should take the lead in preparing these with assistance from the other two-design and O&M.

Initial Complement of Equipment and Supplies. Though a part of POM, a separate document is required to be prepared by the project O&M unit together with the state O&M office describing the necessary complement of fixed and moveable equipment and stock of supplies required at project start-up and at each subsequent stage of project development. This is required to allow timely budgeting, procurement and commissioning before the project services are to commence. Equipment is to include office, shop and field. Supplies are to meet like uses including one-year's spare parts, replacement components (filters, belts, etc.), lubricants, etc. Usually, the project O&M unit is not staffed early enough to do this alone, nor does it have funds to carry these start-up expenditures. It is a necessary part of the initial project investment and must be acknowledged and treated as such. It is assumed that offices, buildings, yards and lands are provided as a part of the usual project construction activity.

Initial Complement of Staff. Though a component of the POM, the initial two-year complement of staff noting specific numbers and qualifications are to be presented in a separate document. Required training courses and method of presentation are to be included. An essential part is the schedule showing recruitment, evaluation, placement, orientation and specific training activities. This must assure a fully capable staff in place at start-up that knows operation and maintenance of equipment, facilities and the systems. Funding and capability constraints necessitate that this also be a part of the initial project activity and investment. This document should be prepared by the state O&M office.

ANNEX 2.1

INDIA
NARMADA SAGAR DAM AND POWER COMPLEX

GENERAL

The capacity of the institutions - organization, regulations, policies and most importantly, the capability of the people - determines the success of an undertaking by Government. The crucial role that the government institutions play in the success or failure of the development and management of the Narmada Basis resources was recognized early in the project formulation. An initial act was the creation of the Narmada Planning Agency (NPA) by GOMP. Assigned responsibility for planning the development of the basin, NPA concentrated on a basin-wide scheme of projects and the more detailed formulation of the Narmada Sagar Complex. Studies went forward on the dams and power plant at the three sites; Narmada Sagar, Omkareshwar and Maheswar, as well as the irrigation projects associated with the first two. This unit directed filed and office work and the preparation of construction documents for early work at the Narmada Sagar. However, as work progressed, the need for expansion and tailoring of the organization to undertake the ongoing planning and execution of the plan became evident. After the receipt of the award (December 1979) of the Narmada Water Dispute Tribunal, which allocated 10.25 MAF waters of the Valley Development was created in June 1981 for the implementation of the multi-purpose major irrigation projects in Narmada Basin.

As an important step, GOMP created a basin authority, the Narmada Valley Development Authority (NVDA) in August 1985, with the responsibility for the planning, development and management of MP's share of the water in the basin and the lands and related resources directly affected. Units have been established to perform the various specialized functions in meeting these responsibilities. The immediate focus of the Authority's activities, which is the basin planning and the execution of the Narmada Sagar Complex, has permitted detailing the organizational arrangements to carry out the component tasks. Staffing plans, position description, staff training and the use of the consultants have been formulated reflecting the work, the organization and the capability of the staff anticipated to be available at the onset. These institutional arrangements are described in detail in this annex.

The Authority was created by government order in July 1985. Patterned on authorities elsewhere in India and on similar organizations in other countries, the entity has jurisdiction for all water resources and related development and management activities within the geographical bounds of the Narmada Basin in MP.

Overall responsibility for the basin activities and the direction of the Authority is vested in the Narmada Control Board (NCB) established by the same order. As a result, the Narmada Control Board will have close review and final approval for essentially all plans and actions proposed by the Authority. The board is chaired by the Chief Minister with the Vice Chairman being the Minister-In-Charge of the Narmada Valley Development Department. Members include the Ministers and Secretaries of GOMP departments involved in activities in the basin such as irrigation, environment, energy, public works, agriculture, forestry, finance and revenue. Other members include the Chairman, and Vice Chairman of NVDA.

Originally the NVDA management consisted of a Chairman; Vice Chairman; Members for planning, engineering, power and finance and ex-officio members which consist essentially of the various Secretaries of line departments involved in the basin. Later, two new members for (i) Rehabilitation and (ii) Environment and Forests were appointed. At present the Deputy Chief Minister of Madhya Pradesh who is the Minister in charge of the Narmada Valley Development Department, is the Chairman of the Narmada Valley Development Authority (NVDA). All members are appointed by the state government and served at its discretion.

Initially, the Chairman of the now abolished Narmada Planning Authority will serve as Chairman of the NVDA. This will assure continuity as the Authority commences operation. Likewise, all staff, functions and assets of the Narmada Planning Authority were absorbed by the Narmada Valley Development Authority .

The order states a number of functions that the Authority will be responsible for. Some of the most pertinent are summarized below:

to prepare a detailed plan for exploitation of the water resources of the Narmada River and its tributaries and to undertake all necessary engineering works for the harnessing of basin waters for the purpose of irrigation, power and navigation and other development;

to undertake ancillary works for the distribution of water for irrigation, industrial, domestic and other purpose;

to undertake generation and sale of power in bulk to MPEB and provision of all necessary engineering works ancillary thereto;

to acquire and manage land in the valley for purposes of carrying out engineering works and provide for human resettlements and other activities to meet the needs for irrigation, flood control and navigation.

to be responsible for human resettlement and rehabilitation and to establish towns and villages and take all necessary measures to ensure planned settlement and rehabilitation;

to advise on the proper conservation and development of forest, wildlife and fisheries in the valley;

to establish a design organization for the projects entrusted to it;

to undertake operation and maintenance of the projects; and

to undertake monitoring and evaluation.

STRUCTURE OF NVDA

The organizational structure of NVDA is depicted in Figure 2.1.1. Offices are shown indicating the reporting responsibilities of the various functions. The Chairman, in addition to overall management of the Authority, has the immediate responsibility for and oversees resettlement and rehabilitation of facilities. The Vice-Chairman, in addition to his other duties, has direct responsibility for the resources council and administration. As may be seen on Figure 2.1.2, oversees the planning program of the Authority. The various specialized functions of the Authority below the Chairman and Vice Chairman come under the direct supervision of the six members, consisting of power, engineering, planning, finance, rehabilitation and environment and forests. The entire power activities from planning to construction and operation and maintenance report to the Member-Power. However, arrangements for coordination with the civil O & M activities, personnel, and maintenance services were refined to assure effective and efficient management and operation of the project. All other planning will be carried out under the Member-Planning, who will receive support from the resources council and the power planning unit. Member-Planning has also been entrusted the work of program monitoring. The Member-Engineering will oversee design and construction of all civil works and minor mechanical/electrical works, pumping plants as well as the operation and maintenance of all Civil Hydro facilities. Member-Finance will oversee the financial functions of the Authority.

As may be noted in Figure 2.1.1, and on the subsequent more detailed charts of the sub-units, earlier it was depicted a program monitoring officer was assigned to a staff position reporting to each unit chief. But program monitoring officers are not specially set out under each Unit-Chief. Program monitoring is entrusted to Member (Planning).

Three important points relative to the program monitoring are emphasized:

All activities of the Authority including those relating to data collection, planning, design, construction, procurement, personnel recruitment, training;

Resettlement and funding which can be described as a series of tasks with completion dates defined and charted. Linkages to other programs were identified on charts at every level;

The program monitoring officer gathers information and present status of programs at weekly intervals and special requests. The status, potential deviations from schedule, and the cause of such deviation was presented to unit managers at regular weekly meetings for their use in making decisions for adjusting staff, securing assisting, or altering the schedule;

Figure 2.1.1   Narmada Control Board

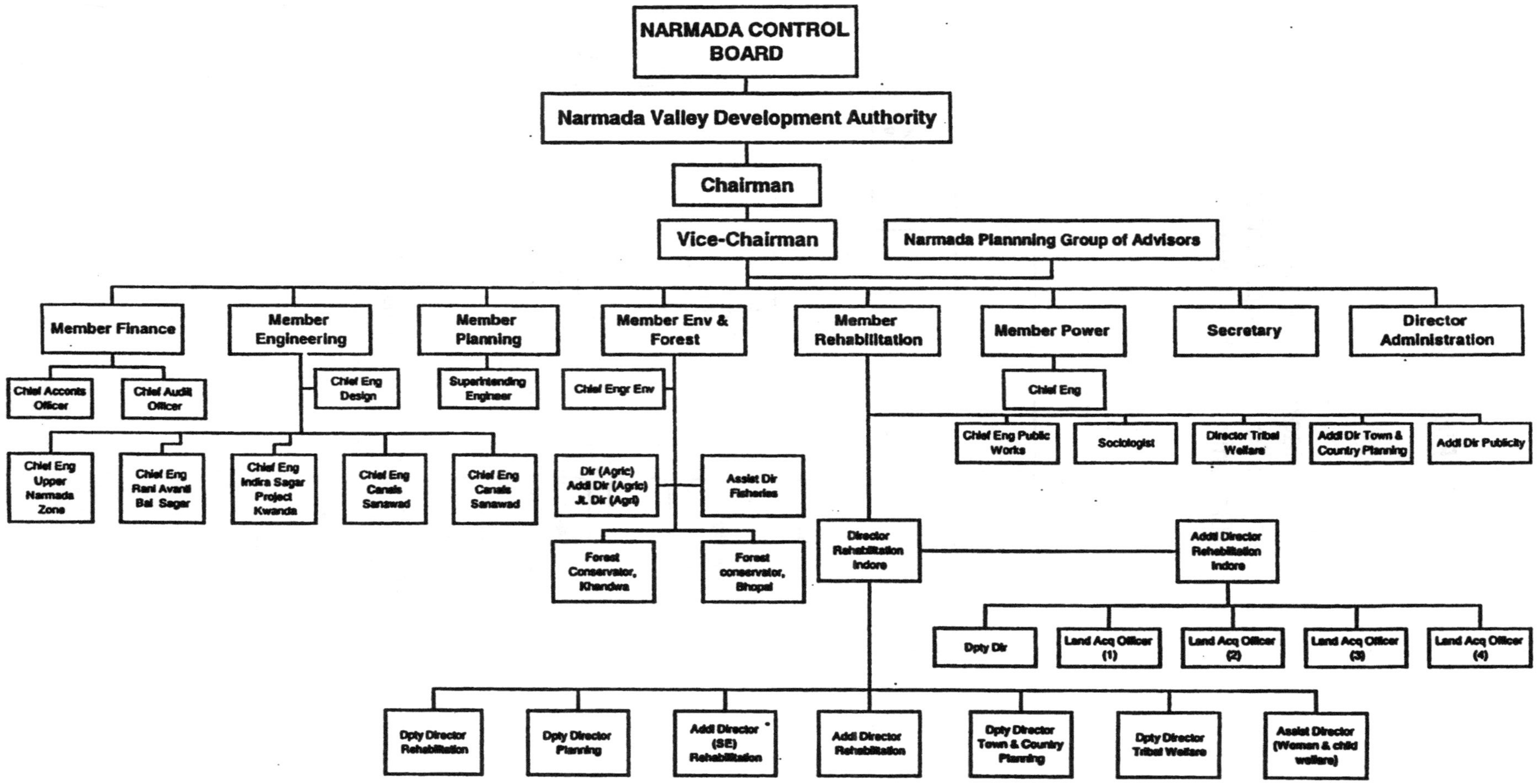

NARMADA CONTROL BOARD

Narmada Valley Development Authority

- Chairman
- Vice-Chairman
- Narmada Planning Group of Advisors

Members:
- Member Finance
- Member Engineering
- Member Planning
- Member Env & Forest
- Member Rehabilitation
- Member Power
- Secretary
- Director Administration

Under Member Finance:
- Chief Accounts Officer
- Chief Audit Officer
- Chief Eng Upper Narmada Zone
- Chief Eng Rani Avanti Bai Sagar

Under Member Engineering:
- Chief Eng Design
- Chief Eng Indira Sagar Project Kwanda
- Chief Eng Canals Sanawad

Under Member Planning:
- Superitending Engineer
- Chief Eng Canals Sanawad

Under Member Env & Forest:
- Chief Engr Env
- Dir (Agric) Addl Dir (Agric) Jt. Dir (Agri)
- Asstd Dir Fisheries
- Forest Conservator, Khandwa
- Forest conservator, Bhopal

Under Member Power:
- Chief Eng
- Chief Eng Public Works
- Sociologist
- Director Tribal Welfare

Under Director Administration:
- Addl Dir Town & Country Planning
- Addl Dir Publicity

Under Member Rehabilitation:
- Director Rehabilitation Indore
- Addl Director Rehabilitation Indore

Under Addl Director Rehabilitation Indore:
- Dpy Dir
- Land Acq Officer (1)
- Land Acq Officer (2)
- Land Acq Officer (3)
- Land Acq Officer (4)
- Dpy Director Town & Country Planning
- Dpy Director Tribal Welfare
- Asstd Director (Women & child welfare)

Under Director Rehabilitation Indore:
- Dpy Director Rehabilitation
- Dpy Director Planning
- Addl Director (SE) Rehabilitation
- Addl Director Rehabilitation

Figure 2.1.2

INDIA
Madhya Pradesh
Narmada Valley Development Authority
(Organization for Planning)

```
                              ┌──────────┐
                              │   Vice   │
                              │ Chairman │
                              └────┬─────┘
          ┌──────────┐             │           ┌──────────┐
          │Resources │─────────────┼───────────│ Program  │
          │ Council  │             │           │Monitoring│
          └──────────┘             │           └──────────┘
```

Member Power	**Member Planning**
	Program Monitoring
Power Planning Unit	**Planning Unit**	**Data Collection Engineering Studies Unit**
Power Studies	Agriculture	Basin Planning	Water Studies
Engineering Studies	Forestry	Major Projects Planning	Engineering Studies
	Fisheries	Medium Minor Project Planning	Hydrology Water Quality
	Demography	Municipal Industrial Waste Treatment	Mapping & Surveys
	Economics		Transportation Studies
	Environment		Geology
	Sociology		

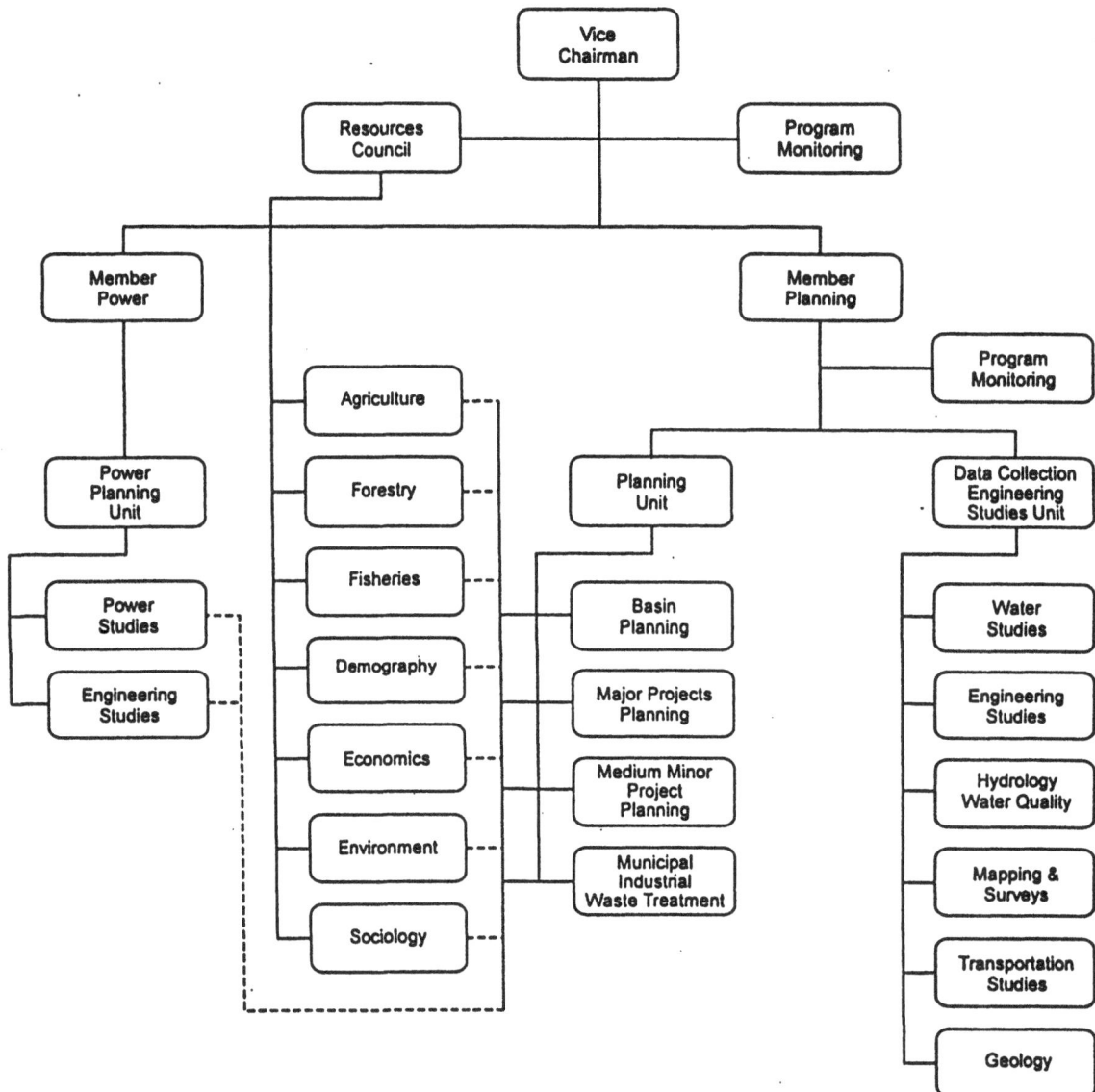

The program monitoring, data storage and retrieval and the presentation of charts was computerized. The selected system permits management at any level to have access to this information on call. This is particularly important once construction commences (first quarter 1986), when decisions have to be made on modifying sequences of work or making adjustments due to unanticipated field conditions. The updated chart of the program monitoring office of each unit was printed and forwarded to the next level of management above, on a routine basis; and

The program monitoring officer exercises no management judgment nor he has any responsibilities relative to decisions on remedial measures or alterations in the programs should delays arise. His sole role is the accurate, prompt reporting of all information and in that regard must be independent of the decision making unit.

It is absolutely essential for efficient, effective direction of the authorities that management at each level and its sub-unit heads expend the effort required to prepare a comprehensive, detailed, accurate program to being with, and that program changes are promptly entered so it is current and accurately presents to management, at every level, the status of all activities. Program monitoring is a critical element to facilitate proper management of the project, particularly recognizing its size, complexity and interrelationships and the time schedule which is of great consequence financially, both from the standpoint of early generation of commercial power and avoidance of delays which incurs extra cost claims by the contractors and increase direct costs of Authority operations.

PLANNING

The Authority has primary responsibility for planning the development of the water resources and the related lands within the boundaries of the Narmada Basin. Irrigation is the dominant consumptive use and hence has the primary impact on both resources. The resulting plan is to guide development through the delineation of the various physical projects and management programs - basin operations, water quality control measures, resources conservation, and the like. The plan is to document the goals and objectives adopted as the basis for the planning. Policies relating to the allocation and use of resources are to be established. The measures for evaluation of projects are to be clearly stated. Priorities of projects and programs are to be set forth.

Planning by its very nature requires updating of the basin plan at intervals as goals change, additional data is secured, and opportunities alter or new ones arise. The plan has to be viewed as a guide reflecting both developments already committed and directions for the future. It must not become a crutch by which past decisions are blindly used as an excuse for undertaking or not modifying a certain project or for following a certain sequence.

The planning function by the Authority is, as a result, an ongoing activity under the direction of the Vice Chairman and the Member-Planning. Figure 2.1.2 illustrates the inputs required form other primary units and indicates areas of expertise contained in the resources council.

The schedule for updating the existing plan was established and given high priority in order to formulate the guidance and have it applied to both present and future activities.

DESIGN

As was shown in Figure 2.1.1, the various areas of expertise and specialized capabilities to carry out the implementation functions of the Authority have been grouped into specific units. These units have basin-wide authority and responsibility and are not limited to individual projects, though initially their sole emphasis is on the Narmada Sagar Complex.

The design unit, headquartered in Bhopal, developsand maintains a staff capability to carry out the obligations of the Authority in the immediate programs for the Narmada Sagar Project, namely, design revisions augmenting the CWC/CEA support to be provided during the construction. It also provides the design capability for dams, tunnels, power plants, pumping plants and canals for the subsequent work, particularly Omkareshwar and Maheshwar which are the next components of the Narmada Sagar Complex.

Figure 2.1.3 presents the units comprising the design organization. Specialization into the categories noted allows concentration and development of the high level expertise necessary and the consistent applications of high standards and most appropriate methods of analysis to all Authority design work. When technical questions arise or questions are posed by the field organization, the best talent in design can be readily identified and brought to bear.

CONSTRUCTION

The construction unit has broad geographical responsibilities, though, as with the design unit, it focuses initially on the Complex. Figure 2.1.1 through 2.1.6 present the organization. A headquarters' office in Bhopal provides a direct link with the design unit on one hand and the operations and maintenance unit on the other. The construction unit's offices in Bhopal are shown on Figure 2.1.4. Its primary function is to effectively manage all construction activities including the dam, the civil portion of the power plant and the canals, as well as any ancillary work involved. It assures the orderly combined procurement of government materials, the prompt delivery of those materials, the uniform treatment of claims and change orders, the review and control of contractors payments and other related matters. An essential role is overseeing and assuring that the field construction staff are properly trained and supported directly or through other means necessary so that quality, cost-effective construction can be assured through prompt and timely supervision and inspection. Another important activity in the Bhopal office of construction is budgeting. This unit prepares quarterly, annual and multi-year budgets reflecting actual and projected contract payments and purchase of government supplied materials. All procurement of materials and works will be centered in the Bhopal headquarters including pre-qualification, preparation of bid documents, bid evaluation and award. Legal experts for contract administration were considered unnecessary.

Beneath the headquarters' office, the next level of organization are the field construction offices. Initially, the principal field office is the one responsible for the construction of the Narmada Sagar dam and power plant. Though these are two large undertakings, they are situated at one site and of necessity require a single field organization for the purposes of management and efficient use of resources and support. This field office has sub-units for the two primary activities - construction supervision of the power plant civil works and of the dam - both of which are serviced by a single office engineering unit to handle quantities, payments, claims, change orders, and materials coordination; a technical support unit providing laboratories and geology; and the administrative assistance unit. Figure 5 presents the arrangement at the Narmada Sagar site.

A like unit, though less complex, having only one type of activity, was established for the canal construction once it is to be launched. These organizations are depicted on Figure 6.

OPERATION & MAINTENANCE

The operation and maintenance organization is depicted in Figure 2.1.7. Basin water operations involving forecasting, basin water allocation and instructions on operation of individual projects in the basin are headquartered in Bhopal. Flood periods require particularly close evaluation and direction of operations by the unit.

Standards, procedures, personnel policies and related activities are also carried out form the headquarters' office.

Units at the next level will be located in the individual projects. The level of staffing of these project offices varies depending on the complexity as can be seen on the chart. The Narmada Sagar Complex entails very major responsibilities and a substantial number of people. The field project office for Narmada Saga Complex O & M consists of the three divisions noted and the various sub-divisions.

Figure 2.1.3
INDIA
Madhya Pradesh
Narmada Valley Development Authority
(Organization for Design)

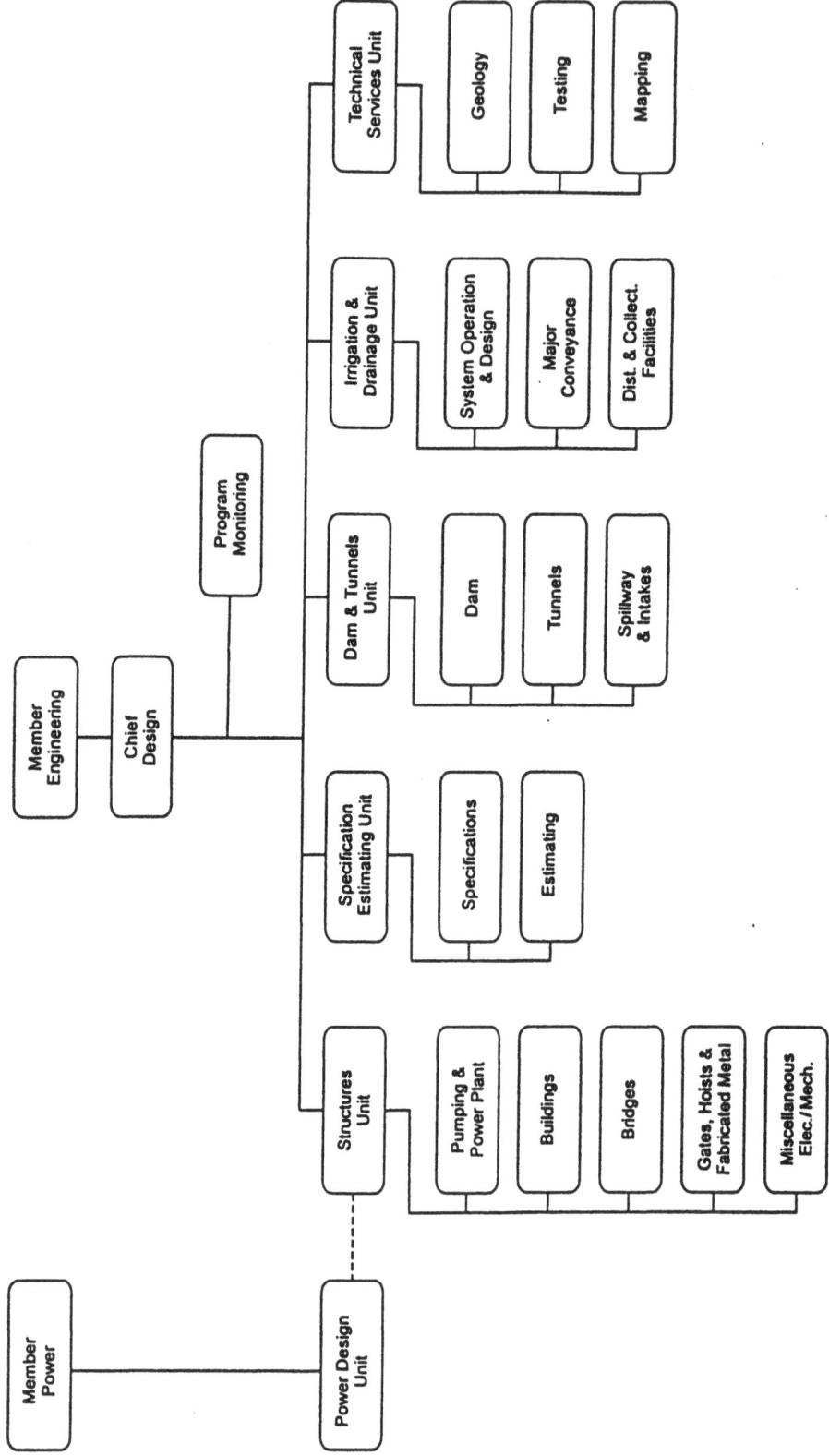

Figure 2.1.4
INDIA
Madhya Pradesh
Narmada Valley Development Authority
(Civil Construction Organization in BHOPAL)

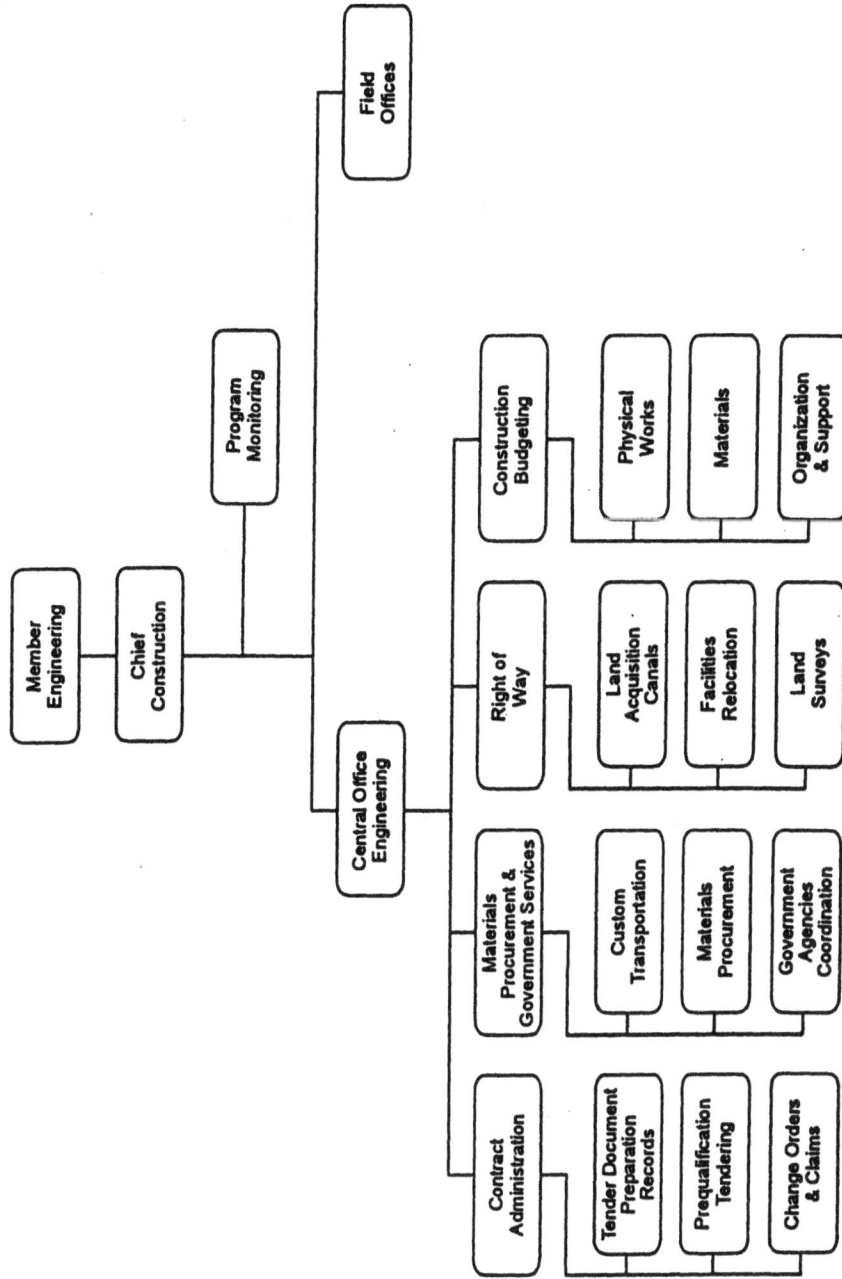

```
                          ┌──────────────┐
                          │    Member    │
                          │  Engineering │
                          └──────┬───────┘
                          ┌──────┴───────┐
                          │    Chief     │
                          │ Construction │
                          └──────┬───────┘
            ┌────────────────────┼────────────────────┐
     ┌──────┴──────┐      ┌───────┴──────┐      ┌──────┴──────┐
     │   Central   │      │   Program    │      │    Field    │
     │   Office    │      │  Monitoring  │      │   Offices   │
     │ Engineering │      └──────────────┘      └─────────────┘
     └──────┬──────┘
```

Contract Administration	Materials Procurement & Government Services	Right of Way	Construction Budgeting
Tender Document Preparation Records	Custom Transportation	Land Acquisition Canals	Physical Works
Prequalification Tendering	Materials Procurement	Facilities Relocation	Materials
Change Orders & Claims	Government Agencies Coordination	Land Surveys	Organization & Support

Figure 2.1.5
INDIA
Madhya Pradesh
Narmada Valley Development Authority
(Construction Organization at Narmada Sagar Dam Site)

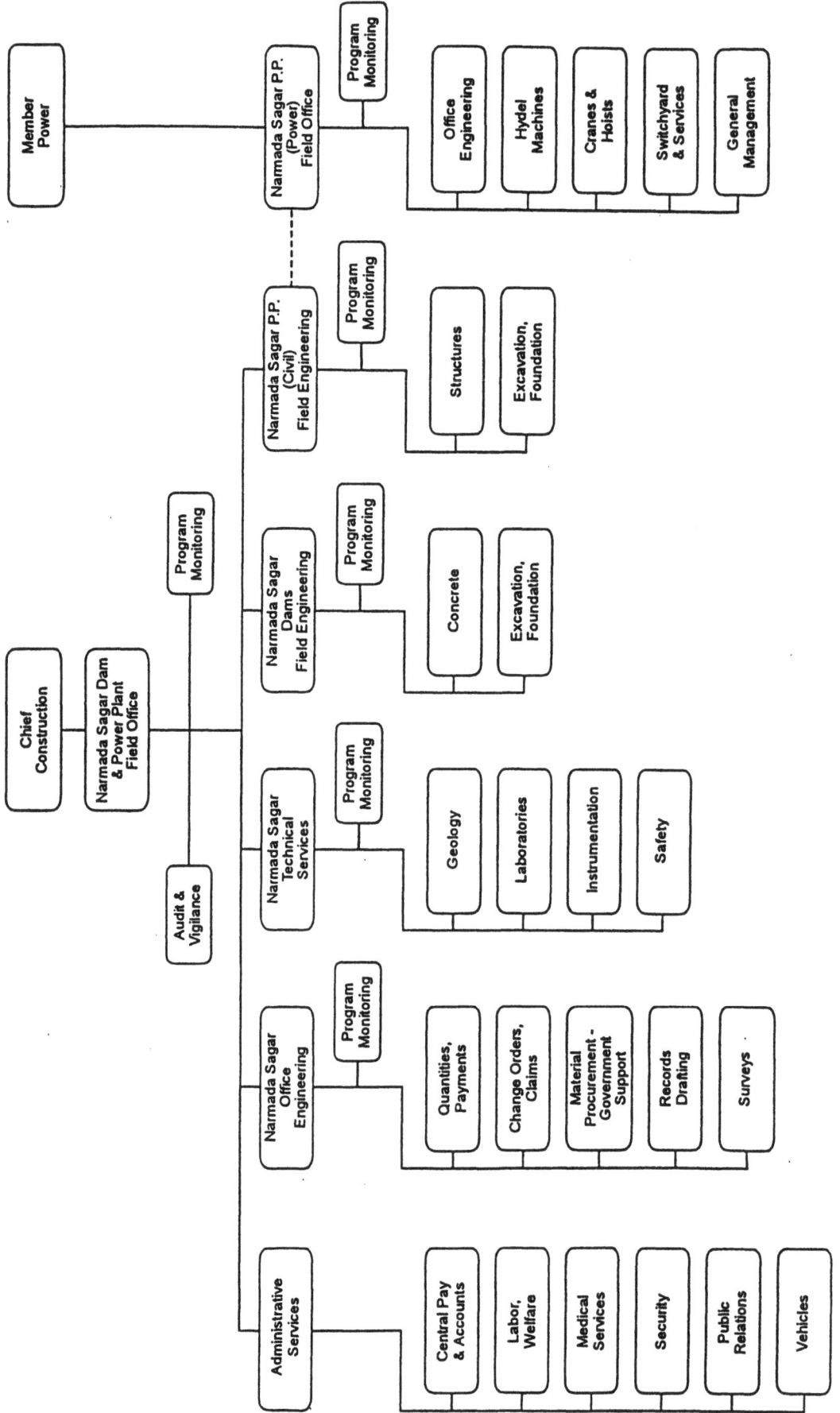

Figure 2.1.6
INDIA
Madhya Pradesh
Narmada Valley Development Authority
(Construction Organization at Irrigation Site)

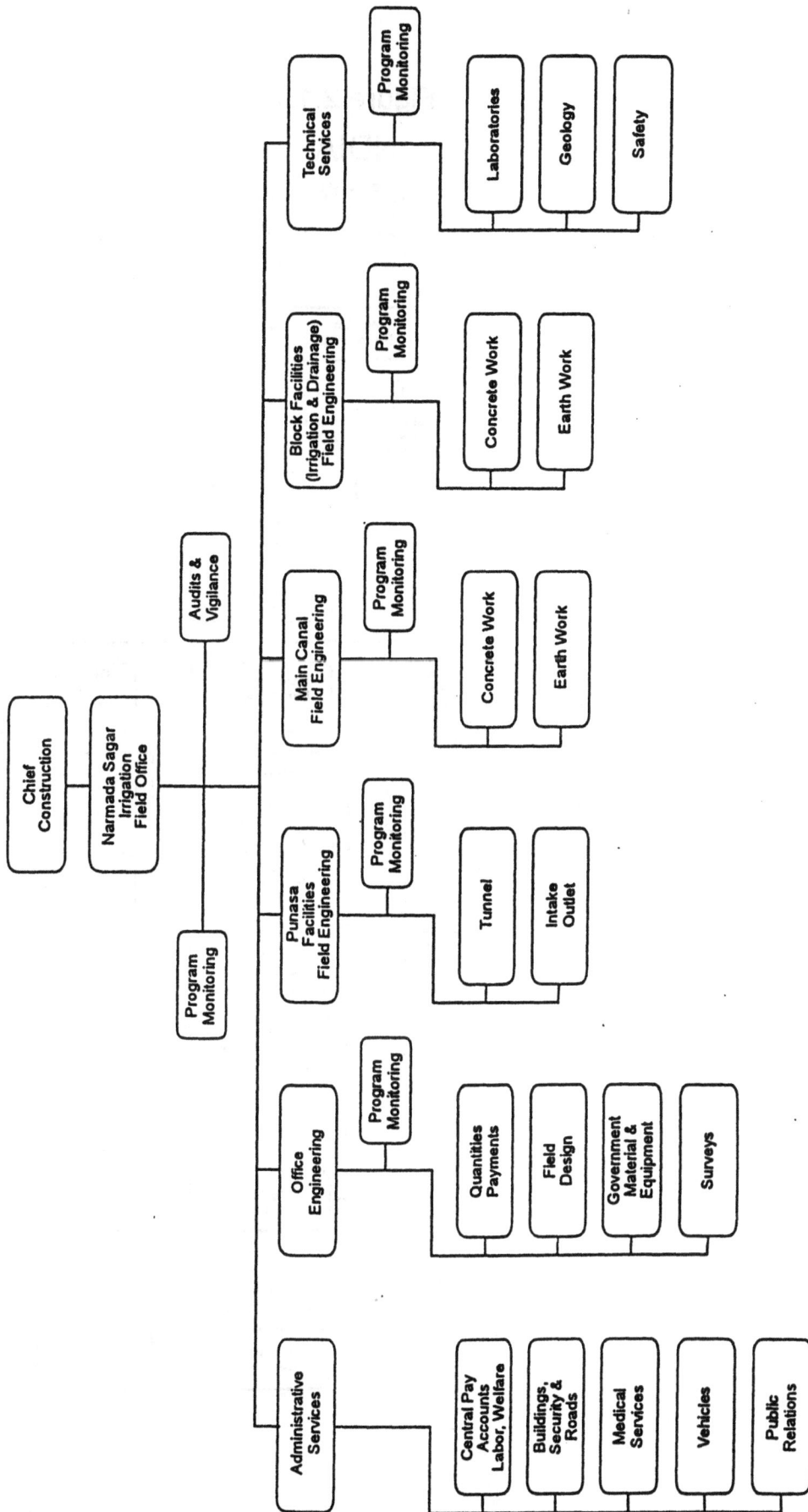

Figure 2.1.7

INDIA

Madhya Pradesh
Narmada Valley Development Authority
(Organization for Operation and Maintenance)

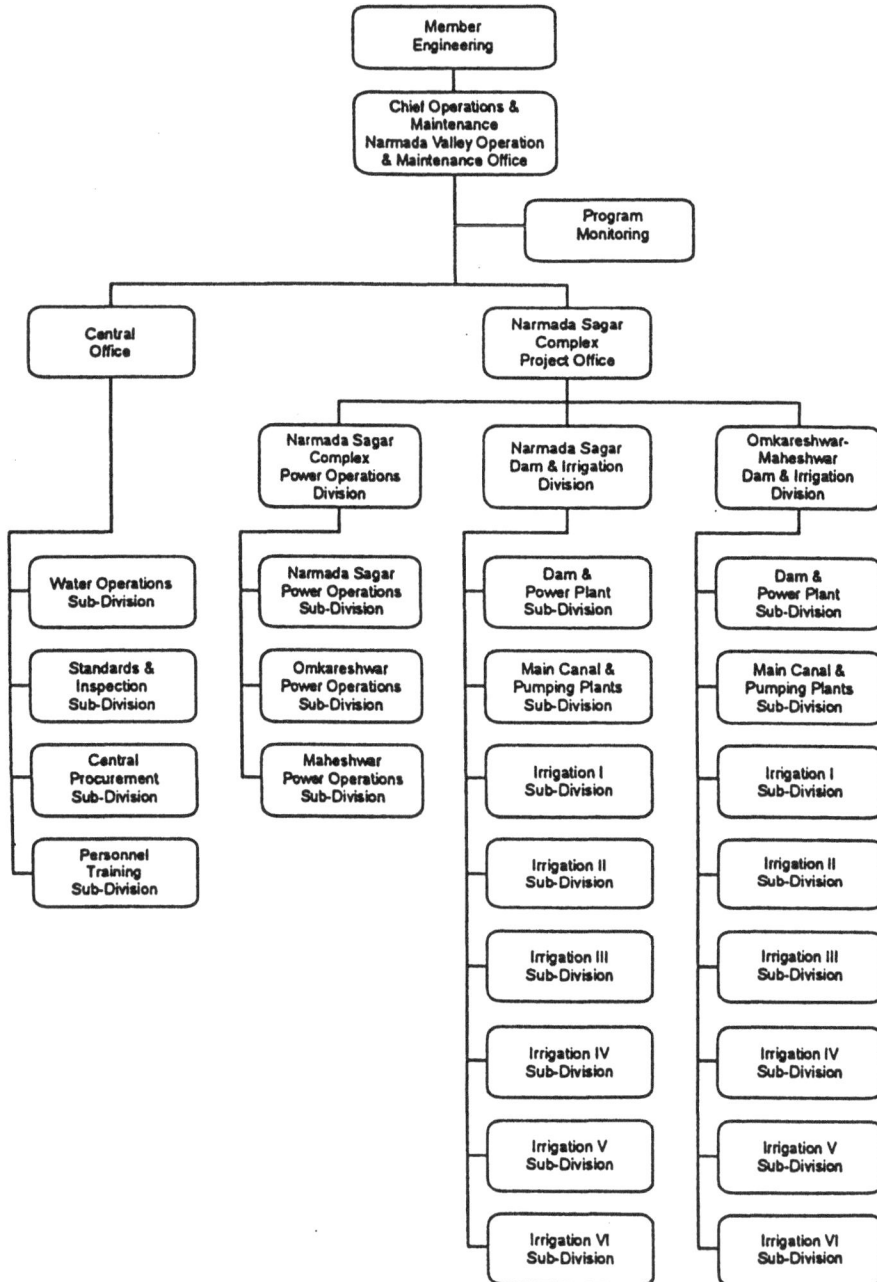

```
                          ┌─────────────────┐
                          │     Member      │
                          │   Engineering   │
                          └────────┬────────┘
                          ┌────────┴─────────────┐
                          │ Chief Operations &   │
                          │     Maintenance      │
                          │ Narmada Valley       │
                          │ Operation &          │
                          │ Maintenance Office   │
                          └──────────┬───────────┘
                                     │        ┌──────────────┐
                                     │        │   Program    │
                                     │        │  Monitoring  │
                                     │        └──────────────┘
         ┌───────────────────────────┴──────────────┐
 ┌───────────────┐                        ┌──────────────────┐
 │    Central    │                        │  Narmada Sagar   │
 │    Office     │                        │    Complex       │
 └───────┬───────┘                        │  Project Office  │
         │                                └────────┬─────────┘
```

Central Office	Narmada Sagar Complex Power Operations Division	Narmada Sagar Dam & Irrigation Division	Omkareshwar-Maheshwar Dam & Irrigation Division
Water Operations Sub-Division	Narmada Sagar Power Operations Sub-Division	Dam & Power Plant Sub-Division	Dam & Power Plant Sub-Division
Standards & Inspection Sub-Division	Omkareshwar Power Operations Sub-Division	Main Canal & Pumping Plants Sub-Division	Main Canal & Pumping Plants Sub-Division
Central Procurement Sub-Division	Maheshwar Power Operations Sub-Division	Irrigation I Sub-Division	Irrigation I Sub-Division
Personnel Training Sub-Division		Irrigation II Sub-Division	Irrigation II Sub-Division
		Irrigation III Sub-Division	Irrigation III Sub-Division
		Irrigation IV Sub-Division	Irrigation IV Sub-Division
		Irrigation V Sub-Division	Irrigation V Sub-Division
		Irrigation VI Sub-Division	Irrigation VI Sub-Division

The initial staffing occurred in the O & M office in commenced in 1986. Refining the operations plan, developing manuals and procedures, undertaking the development of the basin computer model for the purposes of water forecasting and "real-time" operations and importantly undertaking the immediate planning of the O & M facilities to be constructed under the project received first priority. Close coordination was maintained with both the design and construction units to assure a timely, efficient completion of O & M works so that they are available at the time of transfer of responsibility from the construction unit to the O & M unit.

FINANCE

The finance unit is essentially the same as established in other government departments. The Member Finance is the responsible individual in assisting management of the Authority with matters of finance. The units assigned to finance include financial services.

Finance sets out budget guidelines, compile unit budgets and prepare the annual budget of the Authority. Accounting, funding and disbursements are the other primary activities. This unit is responsible for securing and managing funds allocated by the state government for the execution of the Authority's program.

STAFF POSITION DESCRIPTIONS

Staff position descriptions have been prepared for all of the key positions in the planning, design, construction and operation and maintenance units. Position descriptions were prepared by June 1986 for all positions required during the next two years, down through the level of assisting engineers. These statements present the detailed description of the positions in terms of duties, responsibility and authority. Position qualification statements have also been prepared for key positions and likewise will be prepared for the others which present the training, experience and management qualities required of the individuals that are to occupy these positions. Through the use of these two, a clear guide is provided to management for use in filling each slot.

It should be noted that many of the positions demand individuals of substantial and specialized experience and training. Often, many requirements of a position were not met initially by the individual assigned. An important use of the position descriptions and the qualification statements is in comparing these with the credentials of the actual candidate and then determining the training and the consulting support that the individual will require. It was on the basis of these actual needs that the final details and extent of the training program and the hiring of the consultants was based so that they meet the situation as it evolved and that the staff training and assignment of consultants are carried out immediately when that need arises. This is an essential step that has been incorporated into the institutional arrangements for the project.

STAFFING

Schedules of required staff were prepared by unit chiefs and compiled in the personnel unit. Recruitment commenced well in advance of need allowing for change in employer, processing, orientation and training. Provision of additional personnel to allow for staff turn-over and initial inefficiencies was made.

Staff will be recruited mainly from inside, but also from outside the GOMP. Additional sources were considered if vacancies remain unfilled. Evaluation of candidates' abilities was made as applications are received or individuals were approached. The respective unit heads participate with the personnel unit in the evaluation and selection of individuals particularly at the middle and higher levels. The staffing schedule for positions is given in Schedule 2 of the main report.

TRAINING

The training program is refined as staff are selected and required courses are finalized. However, there are certain essential subjects that were identified. These include training in management and personnel that are important in these large units and several technical areas relating to both design and construction. A program has been developed for the initial use in budgeting and planning.

The tentative training program was developed by training specialists from the technical units in the Authority. This permitted incorporation of the views from both the user and the training specialists.

Trainers participating in this program were selected to meet the specific needs e.g. people with long experience in construction management or in field inspection were used as trainers in those subjects rather than professors from universities. At the same time, trainers in management and personnel were selected both from management consulting organizations as well as universities. The technical consulting specialists described in the following section who serve for periods of time in support of both design and construction staff, also dedicate a portion of their time to the training program.

Coordination of the training and updating and refinement of the program as judged best from the results, is under the direction of a full time training officer in the administrative services unit with advice from the supervisors of the respective technical units.

CONSULTANTS

The Central Water Commission (CWC) and the Central Electricity Authority (CEA) in Delhi have been serving as the primary consultants to MP for the design of the dam and power plant facilities in the complex. Their role, as consultant, continued during the construction period of Narmada Sagar dam and power plant. Staff in Delhi, augmented by individuals assigned to Bhopal provide the ongoing service.

It is recognized, however, that the Authority had to support CWC/CEA in producing the 2-3,000 drawings which must be provided in the course of completing these two facilities, and therefore NVDA had to greatly increase its capabilities. Consulting support to the Authority are essential. Consideration was given to using a consulting firm in an overall lead role or in a support role for specific tasks. Consideration was also given to the assignment of consultants to work directly with Authority staff. These, however, would exercise no authority and have no responsibility for actions of NVDA staff. The prime purpose would be the transfer by the Authority and is the contemplated mechanism to be followed, assuming NVDA staffing goals can be met. Accordingly, resident consultant specialist positions have been identified for Bhopal and for the site. Other specialists are to be available on short-term notice to strengthen staff during the initial phases. The design consultants would also help support individuals who will subsequently be involved in the preliminary engineering on the Omkareshwar and Maheshwar facilities.

The primary areas where MP are in need of strengthening their capacity is the management of the construction program. Contract administration, office engineering, field engineering and laboratory testing are examples where specialists were required to augment the Authority staff.

The list of specialists and duration noted in the main report are based on the best present estimates of support required by the Authority staff . The criteria for accepting the individuals was based strictly on their qualifications and experience. Country of residence was not a criterion. This level of consulting support was incorporated into the program initially. However, as Authority staff are assigned to the positions and as the needs alter, this list is modified.

Indeed the method of providing consulting support may have to be altered if the adopted approach does not yield the total capability required for NVDA to fully and efficiently carry out its assignment. Selection of the alternatives of assigning overall responsibility to a consulting firm may ever prove necessary to meet the essential needs. For the presently adopted method, definite advantages exist to have a consulting firm provide the individuals since the firm will assure responsibility for competence and can provide prompt back-up or replacements if needed.

Discussions were also held concerning immediate needs, particularly, that it may be necessary for GOMP to secure a consultant to advise on the detailed structure, procedures, manuals, and forms for the construction management organization. This, however, would be over and above the list of consultants. Likewise, the use of consultants to handle specific tasks in planning or design in order to maintain schedule is also assumed, but is not identified on the list.

ANNEX 2.2

AUSTRALIA
GOULBURN-MURRAY WATER VICTORIA, AUSTRALIA

BACKGROUND

The Goulburn-Murray Rural Water Authority (trading as Goulburn-Murray Water) was created on 1 July, 1994 from the restructuring of the Rural Water Corporation of Victoria, Australia.

On 1 July 1995 the Authority accepted the full accountability for management of State headworks in its region and was appointed the Constructing Authority for Victoria under the Murray-Darling Basin Agreement.

Goulburn-Murray Water's region covers 68,000 square kilometres in the north of the State of Victoria, between the Great Dividing Range and the River Murray. The region includes the major storages and the major gravity irrigation areas in Victoria as well as pumped irrigation and waterworks districts.

GOULBURN-MURRAY WATER'S RESPONSIBILITIES

Goulburn-Murray Water is responsible for :

- management of the major water systems within its boundaries;
- provision of bulk supplies to urban (7) and rural water authorities (2);
- delivery of irrigation water (2.3 million megalitres per annum), domestic and stock supplies, and drainage services to 24,000 serviced properties within the six management areas and along the river systems of northern Victoria.

Goulburn-Murray Water also undertakes a number of natural resource management activities closely related to its core business, for example, salinity and water quality management for Government on a cost recovery basis.

Goulburn-Murray Water's dominant business is the delivery of Rural Water Services, with the dominant line of business comprising irrigation and drainage services.

MISSION STATEMENT

Goulburn-Murray Water will deliver price efficient, sustainable water services to North Victoria.

CORPORATE STRUCTURE

Goulburn-Murray Water has a skills based Board of eight, appointed by the State Minister for Agriculture and Resources and selected for their expertise in a variety of fields including business, finance, engineering, irrigation farming, water systems and environmental management, and includes the Chief Executive.

MANAGEMENT STRUCTURE

The organization is structured into management groups according to their primary functions : Business Development, Headworks, Water Services, Production and Catchments, Corporate Services and Finance.

The business structure reflects the responsibilities and ensures compliance with the Government's desire to separate bulk water services from retail water services, and to ensure the State's valuable water resources are managed in an environmentally sustainable manner.

Through responsible financial management, using a combination of productivity gains, revenue initiatives and price increases, Goulburn-Murray Water will achieve self financing of the real operating and capital cost of each service by 2001 - the date agreed to by the Council of Australian Governments.

ORGANIZATIONAL CHARTS

Charts showing the Goulburn-Murray Water's prime relationships, accountabilities and organizational structure are included, in addition to charts depicting the management structures of Goulburn-Murray Water's six irrigation areas.

FUNCTIONS STATEMENTS

The following sections provide the function statements for the Board, Chief Executive and the six Management Groups, with additional information on the Water Services Group, which has the main responsibility within the organization for operation and maintenance.

Board

The Board determines policy, strategy and financial matters affecting the entire organization. The Board establishes a number of committees which have advisory powers only.

Figure 2.2.1

**GOULBURN-MURRAY WATER (G-MW)
PRIME RELATIONSHIPS**

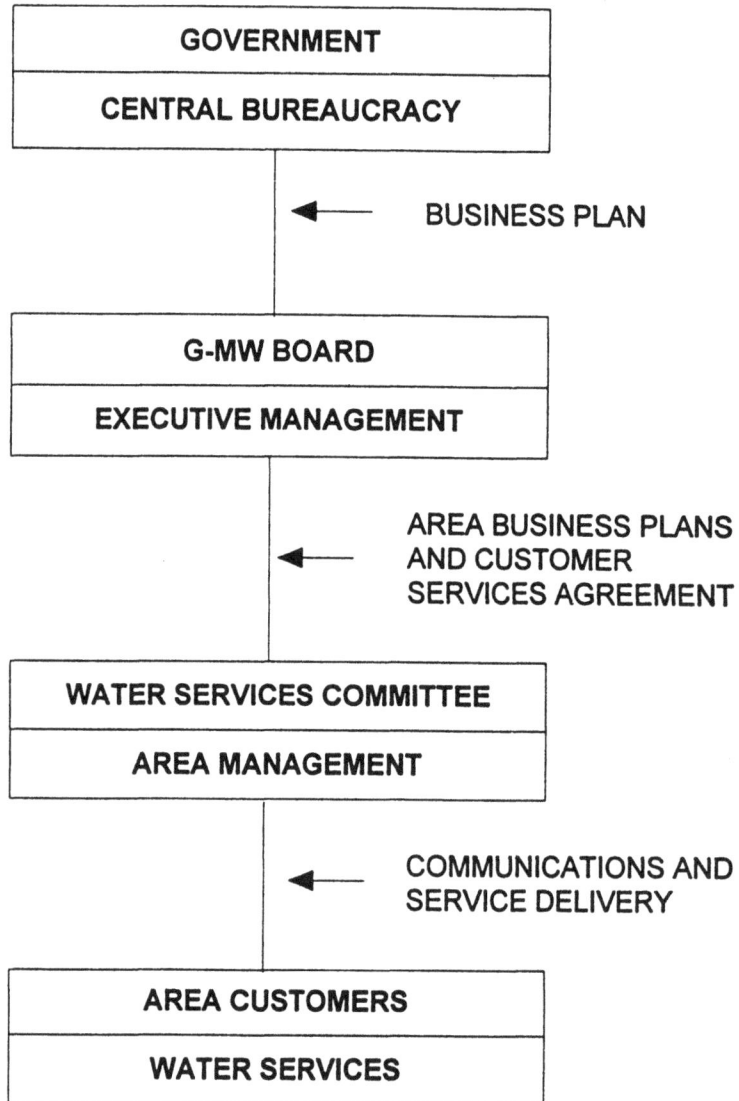

GOVERNMENT
CENTRAL BUREAUCRACY

←—— BUSINESS PLAN

G-MW BOARD
EXECUTIVE MANAGEMENT

←—— AREA BUSINESS PLANS
AND CUSTOMER
SERVICES AGREEMENT

WATER SERVICES COMMITTEE
AREA MANAGEMENT

←—— COMMUNICATIONS AND
SERVICE DELIVERY

AREA CUSTOMERS
WATER SERVICES

Their role is to assist management in the development of policy and strategy, and in monitoring management's implementation of policy and strategy, ensuring that the Board is kept informed.

Chief Executive

The Chief Executive has responsibility for overall management of Goulburn-Murray Water.

Figure 2.2.2

GOULBURN-MURRAY WATER
PRIME ACCOUNTABILITIES

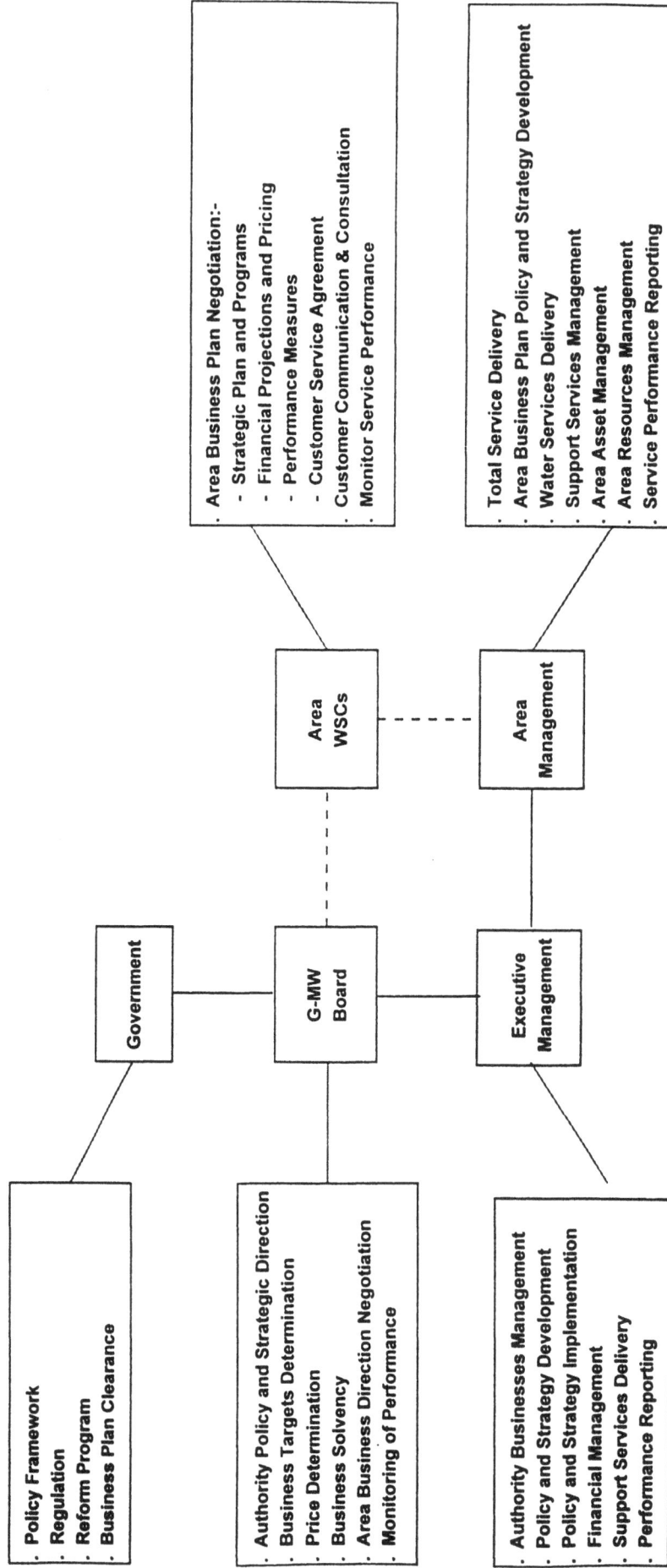

Government

G-MW Board

Executive Management

Area WSCs

Area Management

Policy Framework
. Regulation
. Reform Program
. Business Plan Clearance

Authority Policy and Strategic Direction
. Business Targets Determination
. Price Determination
. Business Solvency
. Area Business Direction Negotiation
. Monitoring of Performance

Authority Businesses Management
. Policy and Strategy Development
. Policy and Strategy Implementation
. Financial Management
. Support Services Delivery
. Performance Reporting

. Area Business Plan Negotiation:-
  - Strategic Plan and Programs
  - Financial Projections and Pricing
  - Performance Measures
  - Customer Service Agreement
. Customer Communication & Consultation
. Monitor Service Performance

. Total Service Delivery
. Area Business Plan Policy and Strategy Development
. Water Services Delivery
. Support Services Management
. Area Asset Management
. Area Resources Management
. Service Performance Reporting

Figure 2.2.3 GOULBURN-MURRAY WATER ORGANISATIONAL STRUCTURE

as of February 1997

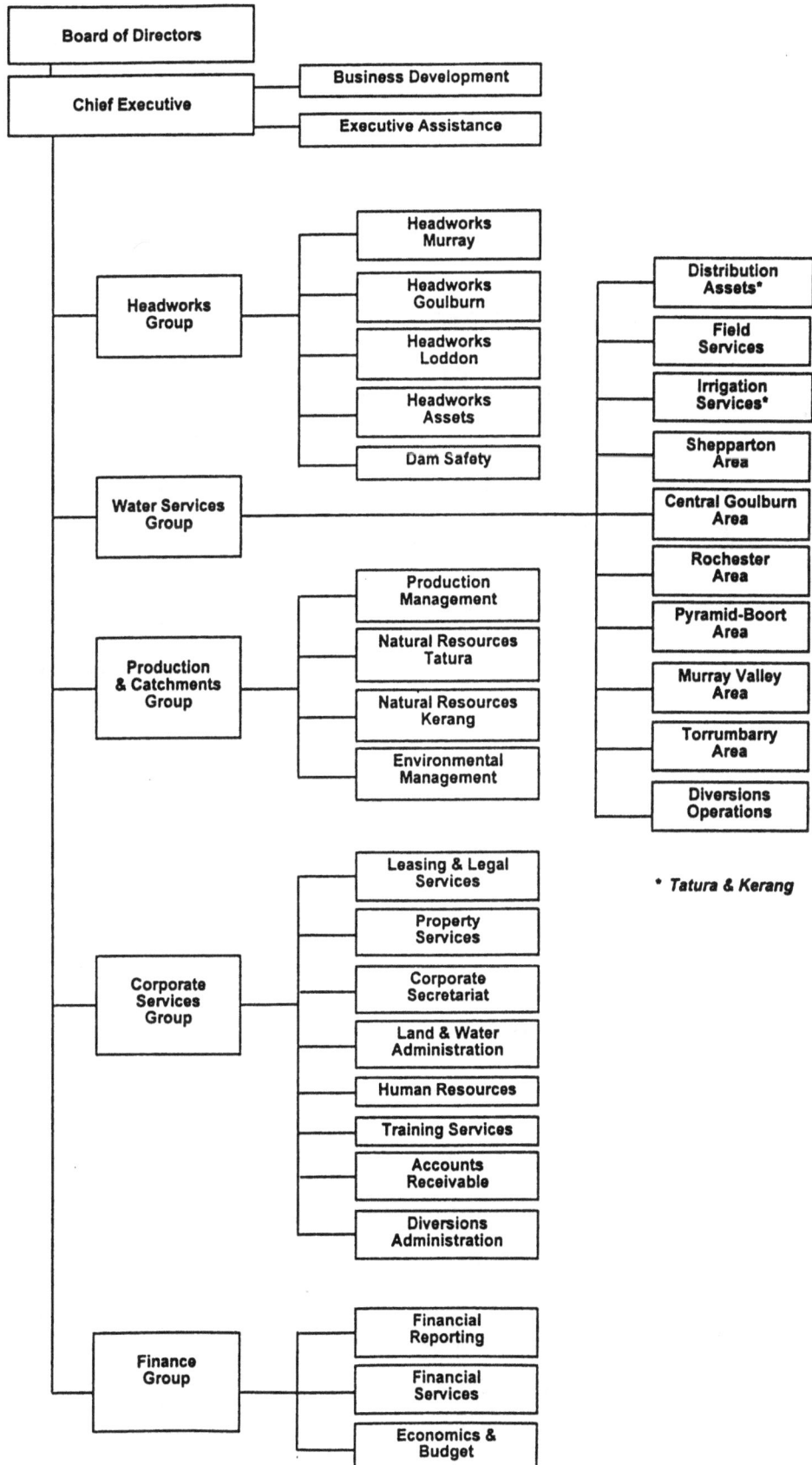

Board of Directors

Chief Executive
- Business Development
- Executive Assistance

Headworks Group
- Headworks Murray
- Headworks Goulburn
- Headworks Loddon
- Headworks Assets
- Dam Safety

Water Services Group
- Distribution Assets*
- Field Services
- Irrigation Services*
- Shepparton Area
- Central Goulburn Area
- Rochester Area
- Pyramid-Boort Area
- Murray Valley Area
- Torrumbarry Area
- Diversions Operations

Production & Catchments Group
- Production Management
- Natural Resources Tatura
- Natural Resources Kerang
- Environmental Management

Corporate Services Group
- Leasing & Legal Services
- Property Services
- Corporate Secretariat
- Land & Water Administration
- Human Resources
- Training Services
- Accounts Receivable
- Diversions Administration

Finance Group
- Financial Reporting
- Financial Services
- Economics & Budget

* Tatura & Kerang

Figure 2.2.4

GOULBURN-MURRAY WATER
TORRUMBARRY AREA MANAGEMENT STRUCTURE
as of February 1997

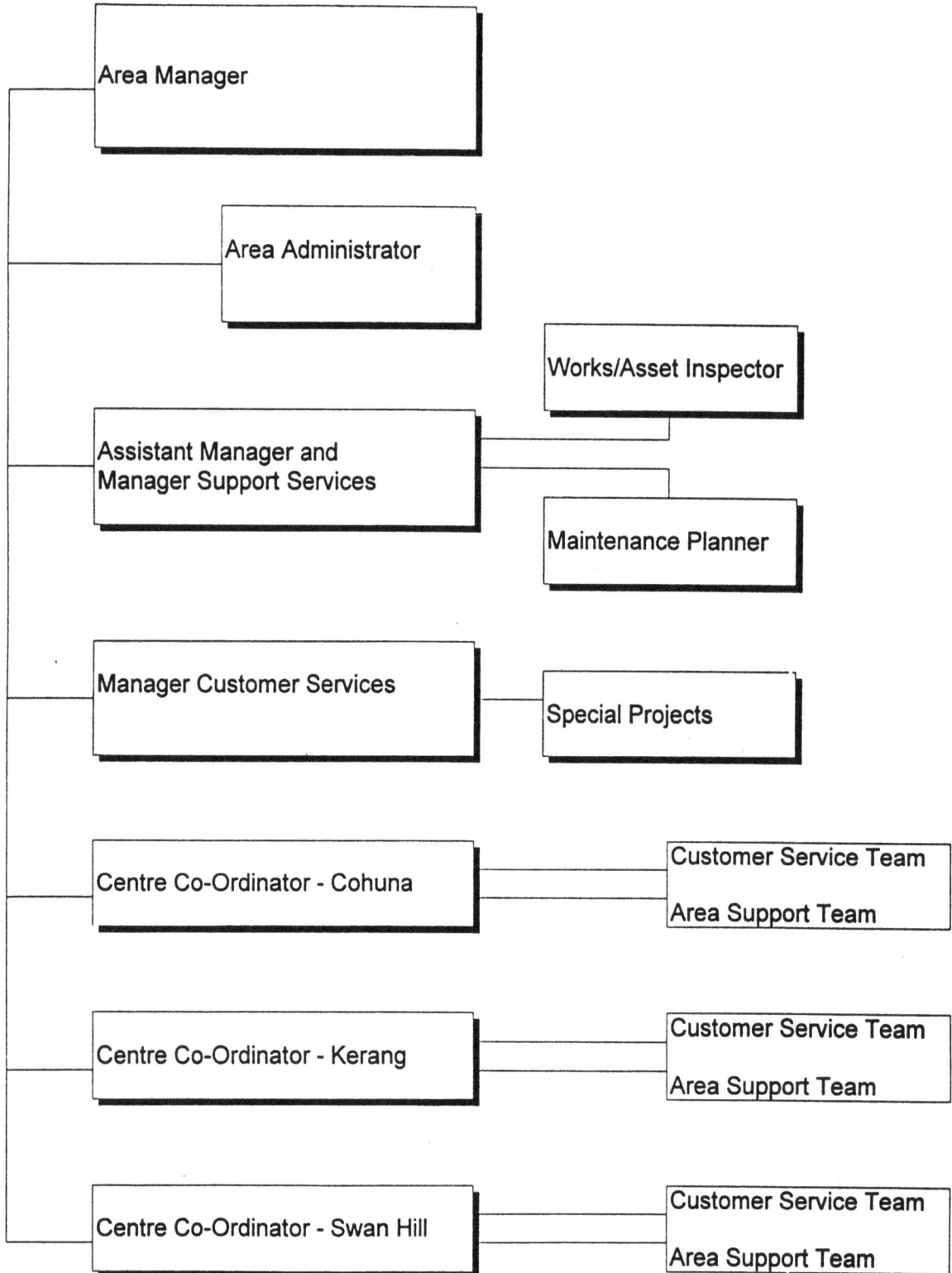

```
Area Manager

Area Administrator

Assistant Manager and          Works/Asset Inspector
Manager Support Services
                               Maintenance Planner

Manager Customer Services      Special Projects

Centre Co-Ordinator - Cohuna   Customer Service Team
                               Area Support Team

Centre Co-Ordinator - Kerang   Customer Service Team
                               Area Support Team

Centre Co-Ordinator - Swan Hill   Customer Service Team
                                  Area Support Team
```

Figure 2.2.5

GOULBURN-MURRAY WATER
SHEPPARTON AREA
MANAGEMENT STRUCTURE
as at February 1997

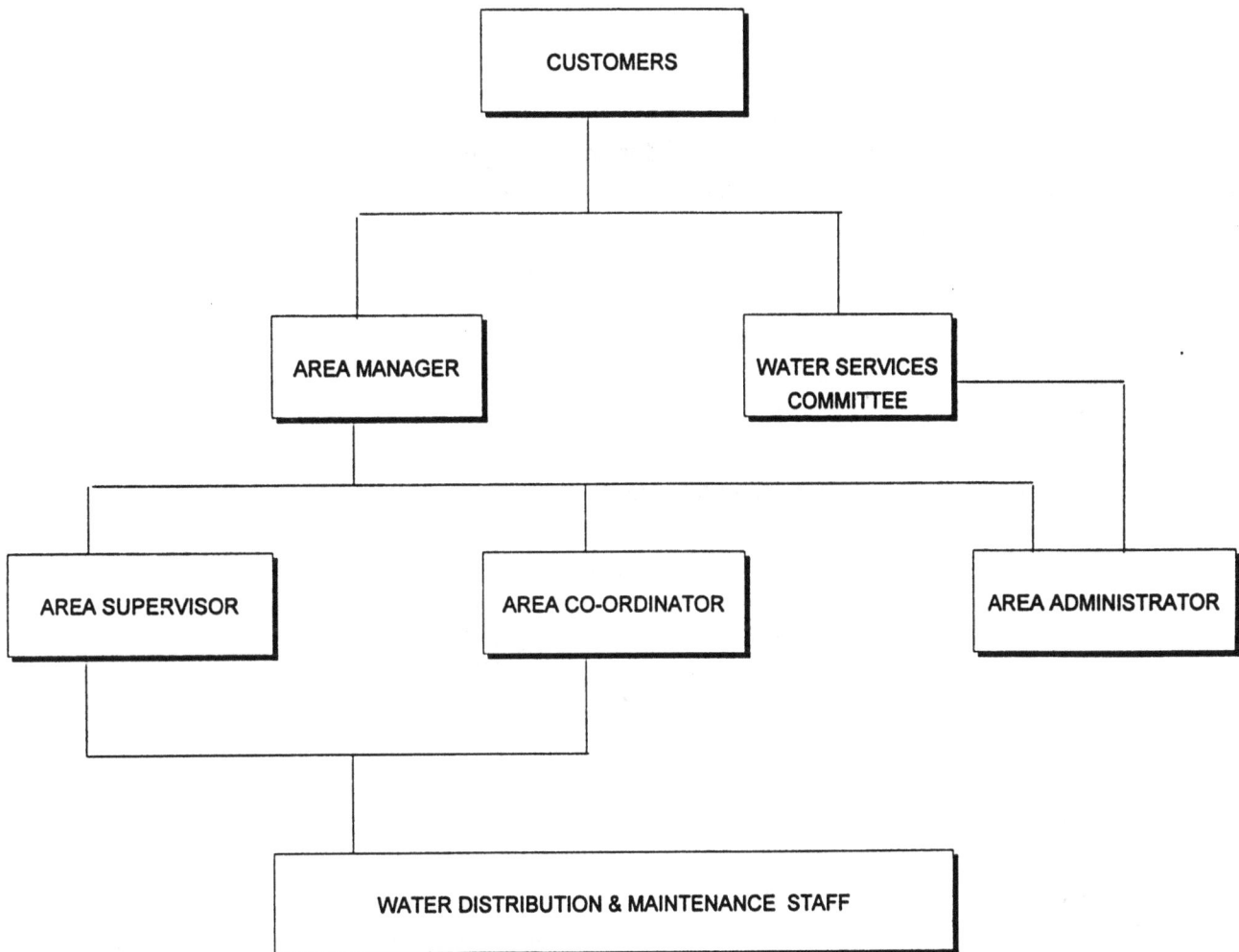

```
                        ┌─────────────────┐
                        │    CUSTOMERS    │
                        └─────────────────┘
                                │
            ┌───────────────────┴───────────────────┐
    ┌───────────────┐                      ┌──────────────────┐
    │  AREA MANAGER │                      │  WATER SERVICES  │
    │               │                      │    COMMITTEE     │
    └───────────────┘                      └──────────────────┘
            │                                                │
  ┌─────────┼──────────────────────────────────────┐        │
┌──────────────┐      ┌──────────────────┐    ┌──────────────────┐
│AREA SUPERVISOR│     │ AREA CO-ORDINATOR │    │ AREA ADMINISTRATOR│
└──────────────┘      └──────────────────┘    └──────────────────┘
      │                        │
      └──────────┬─────────────┘
        ┌────────────────────────────────────────────┐
        │  WATER DISTRIBUTION & MAINTENANCE  STAFF    │
        └────────────────────────────────────────────┘
```

Business Development

The Business Development Group manages Business Planning and Strategic issues for Goulburn-Murray Water.

Headworks

The Headworks Group manages the State headworks within the Goulburn-Murray Water region. The headworks assets provide bulk water, hydroelectricity generation and recreational facilities within the region. Maintenance of asset integrity and the ability to provide secure water supply is of vital importance.

Water Services

The Water Services Group is responsible for the delivery of water and water related services to customers throughout the region, and for the management of distribution assets. Management of Areas and development of Water Services Committees and Customer Groups is a priority. The services delivered to customers by this Group include :

- gravity and pumped irrigation
- private diversions (surface and groundwater diversions)
- domestic and stock supplies
- surface and subsurface drainage
- salinity control
- irrigation technology research and extension

The Water Services Group has "whole of life" responsibility for distribution assets (i.e. those assets used to deliver retail water services to customers). The Group operate, maintain and renew these assets and have technical resources to carry out design and investigations for works in addition to the construction and maintenance resources located in Areas.

Some of the Units within the Water Services Group are :

- Distribution Assets

  The Distribution Assets Unit provides the technical resources (through in-house and consultants) to carry out design and investigations for works and to provide technical inputs, project management and contract management services for the construction and maintenance of distribution assets.

- Irrigation Services

  The Irrigation Services Unit is responsible for ensuring that water services to customers throughout the region are developed to improve the efficiency, profitability and sustainability of irrigation, and that the development of farm irrigation technology better integrates with Goulburn-Murray Water's water systems. The Unit has expertise in all aspects of farm irrigation and drainage practices. An important role of the Unit is to provide internal and external communication of current irrigation and drainage technology and identify changes that will impact on Goulburn-Murray Water operations and opportunities that will lead to more profitable and sustainable irrigation. Development programs are normally based on externally funded research which may be carried out cooperatively with research organizations.

- Field Services

  The Field Services Unit manages major construction projects, production of precast and pre-stressed concrete products; mechanical servicing of fixed and mobile plant and vehicles, and materials testing. The Unit also undertakes construction of private works and works for other Authorities and Departments by agreement. It also provides precast/pre-stressed concrete products and project management services to other Departments and Authorities.

- Irrigation Areas

  The six irrigation Areas are responsible for the delivery of water and water related services to customers throughout the Areas and the operation and management of the distribution assets within Areas. The management structure for each Area reflects the diverse nature of the assets as well as the diverse needs and influences of customer groups. The development of the Water Services Committees and customer groups is a priority.

Water Services Committees

Water Services Committees have been established - six in the irrigation Areas, ten for Diversion customers, and one for Waterworks Districts.

- Water Services Committees represent the customers of Goulburn-Murray Water and play a vital role in defining services, standards of service, maintenance and replacement priorities, and performance measures - all encompassed in a negotiated annual customer service agreement.

- Water Services Committee members maintain contact with the people on the land, and decisions are made in the common interest.

- Decisions made by the Water Services Committees have a major influence on the price of each service.

Production and Catchments

The Production and Catchment Group is responsible for the integrated management of water systems within the whole of the region, and provides a focus for the development and implementation of integrated natural resource management plans and programs including salinity, water quality, landcare and integrated catchment management. The Group is responsible for ensuring the equitable seasonal allocation of water resources and providing direction and advice to the Area Managers on resource availability and utilization.

Corporate Services

The Corporate Services Group is responsible for a range of functions including provision of secretarial and administrative services to the Board; land, water and property management; valuation, property and leasing; administration; accommodation and office needs, information technology; legal and policy advice; human resource management, industrial relations; diversions administration (surface and groundwater); insurance and risk management; training; billing and debt management.

Finance

The Finance Group is responsible for Authority, statutory and management reporting, project and job costing, accounts payable, purchasing and inventory management, treasury management, fixed asset register and asset valuation, financial systems development and maintenance, productivity target negotiations and establishment, budgeting, financial planning, economic evaluations, investment planning and pricing.

ANNEX 3

SUPPLY SERVICES - OBJECTIVES FOR PLACING ORDERS

ESTIMATED VALUE OF REQUIREMENT	OBJECTIVE	COMMENTS
$0 - $1000	5 Working Days	The objective given is for requisitions where contact with suppliers can be made by telephone. Where it is necessary to send out information by mail (e.g., drawings, etc.) lead time could extend to 3 weeks
$1000 - $5000	4 Weeks	Written offers are generally involved in this range to comply with G.S.S.A. guidelines. The lead time is therefore extended to allow written offers to be made. In instances where written offers are not required (e.g. sole supplier) the time required will be shortened.
$5000 - $100,000 (Where assessment is made by Supply Services.)	5 Weeks	These requirements generally require that tenders be advertised. Exceptions can be made where only one supplier exists (e.g., Ductile Iron Pipe). However, if alternatives are to be considered, tenders are required.
$5000 - $100,000 (Where assessment is made by others)	5 Weeks plus assessment time	Where an assessment of offers is to be made by the user and others with an overlapping responsibility, the time taken for this assessment is to be added to standard lead time.
$100,000 - $250,000 (Where assessment is made by Supply Services.)	6 Weeks	The additional time involved is required for assessing, drafting and typing a submission and gaining approval of the Director, Finance and Administration to place an order.
$100,000 - $250,000 (Where assessment is made by others.)	6 Weeks plus assessment time	Refer previous note re: Assessment of Tender by others.
Greater than $250,000 (Where assessment is made by Supply Services.)	8 Weeks	In this instance approval of the General Manager must be obtained and then written sanction of the Executive Council must be sought and obtained prior to placing an order.
Greater than $250,000 (Where assessment is made by others.)	8 Weeks plus assessment time	Refer to previous note re: Assessment of Tender by others.

TYPICAL ORDERING/PURCHASING/RECEIPT PROCEDURE FOR MATERIALS REQUIRED AT A COUNTRY CENTRE

E.g. Supply of RC Pipes
Estimated Value of $50,000

FLOW CHART OF ACTIVITIES

	Activity	Approximate Time Required
1.	Local Inspector identifies need for store requirement from works plan.	1 day
2.	Prepare requisition stipulation exact material requirements including quantity, size, grade, pressure, specification details, desired delivery date, site and any special requirements.	2 days
3.	Submit requisition of Supply Branch.	2 days
4.	Supply Branch initiates purchase by inviting tenders/quotations from suppliers.	3 weeks
5.	Tenders/quotations received and collated by Supply Branch. Tenders evaluated in conjunction with client and technical specialist branch (if necessary).	1 week
6.	Acceptance of best offer arranged by Supply Branch, including statutory approvals as necessary. Place order.	1 day
7.	Supply Branch advises unsuccessful tenderers	1 day
8.	Arrangements made for technical inspection of goods as manufactured at factory prior to delivery.	As required
9.	Local inspector to ensure arrangements made for receipt of goods (site access, receiving office, etc.)	As required
10.	Goods delivered and checked for condition and compliance with order.	As stated in order
11.	Goods receipted and authority issued by receiving officer for payment in accordance with terms of order.	1 day

ANNEX 4

Program for Development
of 1987 Corporate plan and 1987/88 Budget

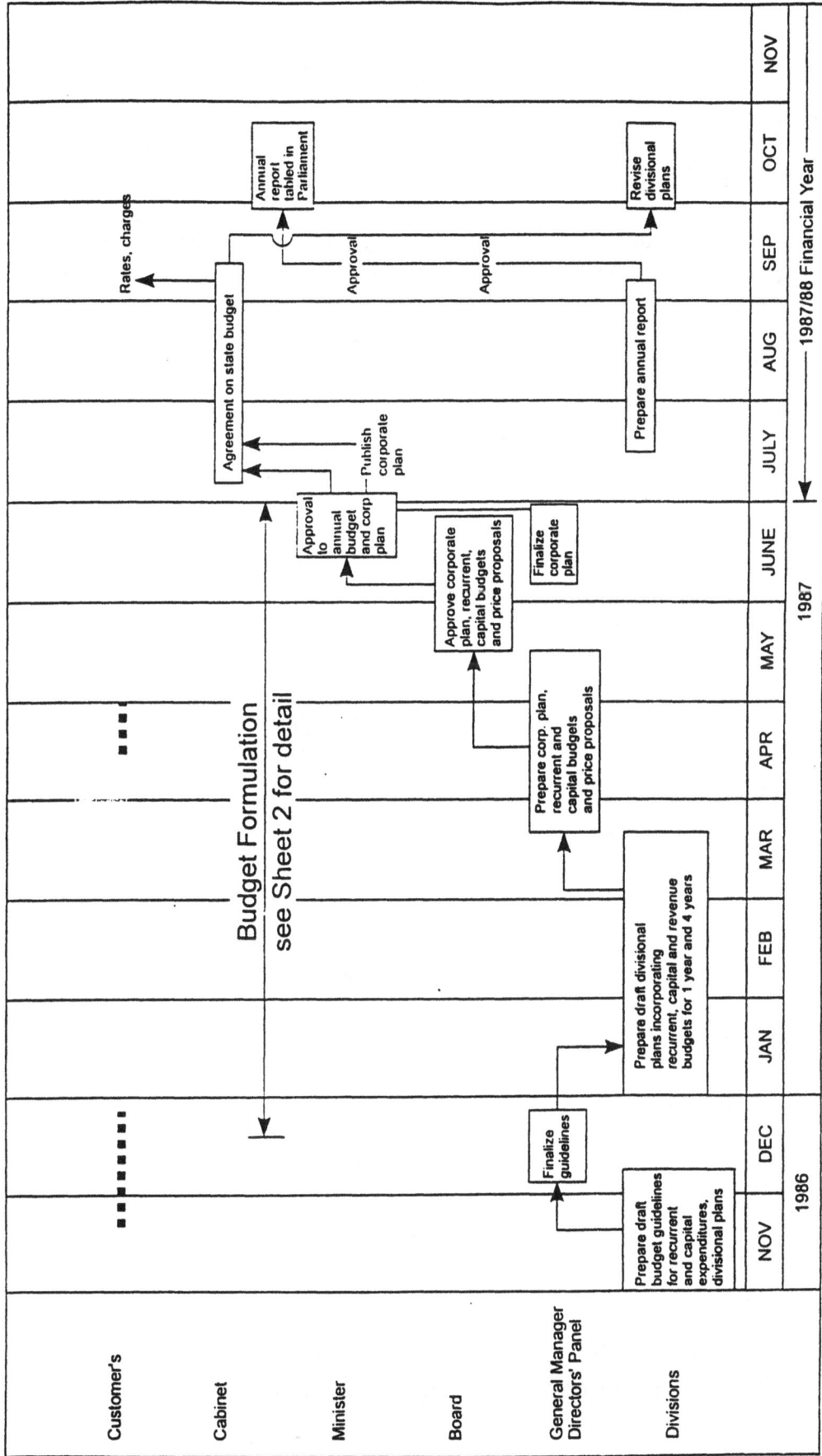

	NOV	DEC	JAN	FEB	MAR	APR	MAY	JUNE	JULY	AUG	SEP	OCT	NOV

Customer's

Cabinet — Rates, charges

Minister — Annual report tabled in P'arliament; Agreement on state budget; Approval; Approval

Board — Approval to annual budget and corp plan; Publish corporate plan; Approve corporate plan, recurrent, capital budgets and price proposals; Finalize corporate plan; Revise divisional plans

General Manager Directors' Panel — Finalize guidelines; Prepare corp. plan, recurrent and capital budgets and price proposals; Prepare annual report

Divisions — Prepare draft budget guidelines for recurrent and capital expenditures, divisional plans; Prepare draft divisional plans incorporating recurrent, capital and revenue budgets for 1 year and 4 years

Budget Formulation see Sheet 2 for detail

1986 | 1987 | 1987/88 Financial Year

■ ■ ■ Key consultative phases

1987/88 BUDGET TIMETABLE

Note: Timetable anticipates DMB budget timetable. Any changes resulting from release of DMB timetable will be incorporated when available.

09/01/87   Distribution of 1987/88 budget package including General Manager budget guidelines and budget documentation to RA Managers and FE Managers as applicable. Budget package will include requirement to identify recurrent expenditure and capital expenditure projections per component for 1988/89 and 1989/90.

13/02/87   Consultative process between Service Units, RA and FE Managers completed. RA's to return relevant form to DFA identifying service unit requirements for 1987/88.

20/02/87   Revised 1986/87 Budget (recurrent, Capital and Revenue) and forward look estimates on unchanged policy basis provided to Department of Management and Budget and Minister.

27/02/87   Budget package completed including initiatives and returned by RA Managers to directors. Directors to review budgets in conjunction with FE Managers and approve the estimates for inclusion in Draft commission budget for 1987/88 to be finalized on 14/5/87.

                   Copy of estimates provided to Finance and Administration Division for entry to General Ledger.

                   Revenue budget estimates options per financial entity prepared by Finance and Administration Division for consideration by FE Managers.

06/03/87   Budget package as approved by each Director provided to General Manager and Finance and Administration Division for correction (if any) to estimates entered the previous week into General Ledger by Finance and Administration Division.

                   Estimates approved by each Director provided to Corporate Planning for review of priorities.

13/03/87   General Manager to review RA budgets and approve their inclusion in the Draft Commission Budget for consideration at the Board meeting on 15/5/87.

                   Following approval by General Manager Finance and Administration Division commence preparation of Draft Commission budget.

03/04/87   Revenue budget estimates as amended if required by FE Managers returned to Finance and Administration Division from applicable Directors.

06/04/87   All Advisory Boards and/or Peak Councils to meet during the period commencing 13/4/87 to 1/5/87 and recommend to the Commission and Government, rates of return, indicative price levels, subsidy levels (if any), recurrent and capital expenditure levels for 1987/88 and projected s for 1988/89 and 1989/90.

27/04/87   Draft Commission Budget for 1987/88 including initiatives considered by General Manager and Directors' Panel. Directors to present to the General Manager RA and FE budget for those areas within their responsibility. One revenue option to be chosen for presentation of profit and loss, balance sheet and cash statements per business FE and for the Commission in total. Impact of alternatives will be shown on financial models.

14/05/87   Commission Budget for 1987/88 and financial plan 1987/88 - 1989/90 including consumer group recommendations presented to Board for consideration and guidance. Draft Budget provided by Board to Minister for advice.

05/06/87   Revised Budget presented to General Manager for approval (incorporating any changes resulting from consultations with consumers, revised priorities, board and Ministerial advice).

13/06/87   Policy initiatives for 1987/88 provided to Department of Management and Budget for consideration by the Priorities Planning and Strategy Committee of Cabinet.

1987/88   Program Structure and narrative provided to Department of Management and Budget.

18/06/87   Budget presented to Board for approval.

30/06/87   Budget presented to Minister for approval.

                Budget provided to Department of Management and Budget for inclusion in first draft State Budget.

July-   Discussions and negotiations with the Department of Management and
August   Budget on details of the 1987/88 Budget and initiatives.

REFERENCES

1.   Alberta Agriculture. 1980. "Post-Construction Management of Drainage Systems," Agric-fax No. Agdex 752-6, Edmonton, Alberta.

2.   American Society of Agricultural Engineers. 1985. "Drip/Trickle Irrigation in Action," Proceedings of the Third International Drip/Trickle Irrigation Congress, Nov. 18-21, St. Joseph, Michigan: ASAE.

3.   American Society of Civil Engineers. 1980. "Operation and Maintenance of Irrigation and Drainage Systems," ASCE, Manuals and Reports on Engineering Practice No. 57, New York.

4.   Arceneaux, W. 1974. "Operation and Maintenance of Wells." *Journal of the American Water Works Association.* March: 199-204.

5.   Ayres, C.E., and D. Scoates. 1928. *"Land Drainage and Reclamation,"* New York: McGraw-Hill.

6.   Bennison, E.W. 1953. "Fundamentals of Water Well Operation and Maintenance." *Journal of the American Water Works Association,* March: 252-258.

7.   Booher, L.J. 1974. *"Surface Irrigation."* Rome, Italy: Food and Agriculture Organization of the United Nations.

8.   Borchelt, J.G., ed. 1982. "Masonry: Materials, Properties and Performance." ASTM Special Technical Publication 778, Philadelphia.

9.   Brick and Tile Institute of Ontario. 1965. *"Clay Masonry Manual."* Willowdale, Ontario, Canada.

10.   Bucks, D.A., et al. 1979. "Trickle Irrigation Water Quality and Preventive Maintenance." Agricultural Water Management.

11.   Campbell, M.D., and J.H. Leher. 1973. *"Water Well Technology."* New York: National Water Well Association, McGraw-Hill.

12.   Central Board of Irrigation and Power, Government of India. 1956. "Investigation Manual for Storage Reservoirs." Publication No. 58, Central Electric Press, New Delhi, India.

13.   Central Board of Irrigation and Power. 1980. "Proceedings of the Symposium on Operation and Maintenance of Canal Systems - May 2-3, 1980," Publication No. 144, Central Electric Press, New Delhi, India.

14. Central Board of Irrigation and Power. 1967. "Symposium on Canal Lining." Publication No. 82, Central Electric Press, New Delhi, India.

15. Central Board of Irrigation and Power. 1977. "Symposium on Silting and Reservoirs with Special Reference to Estimating the Life Reservoirs and Measures to Arrest the Rate of Sedimentation," Publication No. 126, Central Electric Press, New Delhi, India.

16. Coote, D.R., et al. 1984. "Reducing Erosion of Open Channel Drains in Problem Soils of the Ottawa, St. Lawrence Lowlands of Canada," International Commission on Irrigation and Drainage, 12th Congress, Fort Collins, Central Electric Press, New Delhi, India.

17. Cox, R.A. 1983. *"Technicians Guide to Programmable Controllers,"* Delmar Publishers, Albany, New York, USA.

18. Davis, G.B. 1964. *"An Introduction to Electronic Computers,"* McGraw-Hill, New York, USA.

19. Doneen, L.D., and D.W. Westcot. 1984. "Irrigation Practice and Water Management," Paper 1 - Rev.1, Food and Agricultural Organization of the United Nations, Rome, Italy.

20. Economic Commission for Asia and the Far East, Bangkok, Thailand. 1968. "Proceedings of the Eighth Session of the Regional Conference on Water Resources Development in Asia and the Far East," November: Water Resources Series No. 38, United Nations, New York, USA.

21. Fausay, N.R., et al. 1982. *"Subsurface Drain Maintenance in Ohio,"* American Society of Agricultural Engineers, Transactions, St. Joseph, Michigan, USA.

22. Gilbert, R.G., et al. 1979. *"Trickle Irrigation: Prevention of Clogging,"* Transactions of the American Society of Agricultural Engineers, St. Joseph, Michigan.

23. Gilles, K.P. "Aspects of Optimization of Canal System Maintenance," International Commission on Irrigation and Drainage, 10th Congress, Athens, Central Electric Press, New Delhi, India.

24. Grass, L.B., et al. 1975. "Inspecting and Cleaning Subsurface Drain Systems," United States Department of Agriculture, Agricultural Research Service, United States Government Printing Office, Washington, D.C.

25. Hansen, V.E., and O.W. Israelsen. 1980. *"Irrigation Principles and Practices,"* 4th ed., John Wiley and Sons Inc., New York, USA.

26. Hodgins, B. 1977. "Installation, Operation and Maintenance of Vertical Turbine Pumps, Water and Pollution Control."

27. Hillel, D., ed. 1982. *"Advances in Irrigation,"* Vol. 1, Academic Press, New York, USA.

28. Hill, R.A. 1950. "Operation and Maintenance of Irrigation Systems," Transactions, American Society of Engineers, USA, Paper No. 2480, December: 72-79.

29. Houk, I.E., 1956. *"Irrigation Engineering,"* Vol. II, John Wiley and Sons Inc., New York, USA.

30. Hubert, C.I. 1955. *"Preventive Maintenance of Electrical Equipment,"* McGraw-Hill, New York, USA.

31. Hussain, M.I. 1981. "Appropriate Construction and Maintenance Technology for Irrigation and Drainage Works in Developing Countries," State-of-the-Art Irrigation, Drainage and Flood Control No. 2, edited by K.K. Framji, International Commission on Irrigation and Drainage, Central Electric Press, New Delhi, India.

32. International Commission on Irrigation and Drainage, Australian National Committee. 1966. "Recent Awareness in Maintenance of Irrigation Channels and Drains," Annual Bulletin, Central Electric Press, New Delhi.

33. Jansen, R.B. 1980. "Dams and Public Safety," U.S. Dept. of the Interior, Water and Power Service, United States Government Printing Office, Denver, Colorado.

34. Jenson, M.E., ed. 1980. "Design and Operation of Farm Irrigation Systems," American Society of Agricultural Engineers, Monograph No. 3, St. Joseph, Michigan, USA.

35. Jones, C.W. 1983. "Frost Induced Slides on Membrane-Lined Canal Slopes," Civil Engineering, ASCE, New York, USA. November: 68-69.

36. Jones, C.W. 1983. *"Performance of Granular Soil Covers on Canal Linings,"* Journal of the ASCE, Irrigation and Drainage Division, New York, USA.

37. Jones, L.D. 1976. "Irrigation Systems, Part 2: Their Care and Maintenance," The Dairyman, Victoria, Australia, December: 11-13.

38. Khosla, A.N. 1953 "Silting of Reservoirs," Central Water and Power Commission, Government of Indian Press, Simla, India.

39. Kraatz, D.B. 1977. *"Irrigation Canal Lining,"* Land and Water Development Series No. 1, Food and Agriculture Organization of the United Nations, Rome, Italy.

40. Mills, H.J. 1971. *"Operation and Maintenance of the Colorado River Aqueduct,"* Journal of the ASCE, Irrigation and Drainage Division, New York, USA. March: 203-209.

41. Mohanty, R.B., and P.K. Misra. 1972. "Design, Construction and Maintenance of the Agricultural Drainage Systems," Proceedings of the Symposium on Waterlogging, Causes and Measures for Its Prevention, Publication No. 118, Central Board of Irrigation and Power, Central Electric Press, New Delhi, India. December: 123-137.

42. Nelson, M.L., et al. 1970. "Report on Potential Growth of Aquatic Plants of the Lower Mekong River Basin, Laos, Thailand," U.S. Agency for International Development, Corps. of Engineers, Washington, D.C.

43. North Dakota State University, Cooperative Extension Service, "Irrigation Hand Book, Operation and Maintenance of Irrigation Wells, Circular AE-97," Fargo, North Dakota.

44. Ontario Ministry of Agriculture and Food, "Drainage Guide for Ontario," Publication 29, Toronto, Canada.

45. Pair, C.H., ed. 1983. "Irrigation" 5th Edition, The Irrigation Association, Silver Spring, Maryland.

46. Pair, C.H. 1966. "Sprinkler Irrigation," United States Department of Agriculture, Agricultural Research Division, Washington, D.C. November: (Revised).

47. Peters, N. and W.C. Long. 1981. "Performance Monitoring of Dams in Western Canada," Amerian Society of Civil Engineers 1981 International Convention. May: 11-15.

48. Ploss, L.F. 1983. "Maintenance Standards for a Distribution System with a Limited Water Supply," Proceedings of the Specialty Conference on Advances in Irrigation and Drainage, Surviving External Pressure, American Society of Civil Engineers, Irrigation and Drainage Division, New York, USA. July: 154-159.

49. Prairie Agricultural Machinery Institute. 1984. "Evaluation Report No. 388 - Lockwood Model 2265 Central Pivot Irrigation System with Flexspan Corner System Attachment," Lethbridge, Alberta, Canada.

50. Prairie Agricultural Machinery Institute. 1984. "Evaluation Report No. 348, Valley Universal Rainger Model 9880 Linear Move Irrigation System," Lethbridge, Alberta, Canada.

51. Prairie Farm Rehabilitation Administration. "Upstream Slope Protection for Earth Dams in the Prairie Provinces," Regina, Saskatchewan, Canada.

52. Ralston, A., ed. 1983. "Encyclopedia of Computer Science and Engineering," Second Edition, Van Nostrand Reinhold, New York.

53. Ridinger, R.D., and C.R. Burrows. "Operation and Maintenance of Automatically Controlled Pumping Plants Providing Pressure for Large-Scale Sprinkler Irrigation

Systems," International Commission on Irrigation and Drainage, 8th Congress, Varna, Central Electric Press, New Delhi, India.

54. Rolland, L. 1982. "Mechanical Sprinkler Irrigation," Paper 35, Food and Agriculture Organization of the United Nations, Rome, Italy.

55. Sagardoy, J.A., et al. 1982. "Organization, Operation, and Maintenance of Irrigation Schemes," Food and Agriculture Organization of the United Nations, Rome, Italy.

56. Shady, A.M., and R.S. Broughton. 1976. "Maintenance and Checking of Performance of Subsurface Drainage Systems," McGill University, Department of Agricultural Engineering, Ste. Anne de Bellevue, Quebec, Canada.

57. Shanklin, D.W. 1984. "Repair of Concrete Water Resource Structures by Epoxy Materials," Paper No. 84-2643, American Society of Agricultural Engineers, St. Joseph, Michigan, USA, December 11-14.

58. Templeton, H.C. 1971. "Valve Installation, Operation and Maintenance," Chemical Engineering, Deskbook Issue, October 11.

59. Theissen, J., and Smith. 1982. *"Modernizing Irrigation Systems in Alberta,"* Canadian Journal of Civil Engineering, Vol. 9, No. 2, Canada.

60. United States Army Corps of Engineers. 1970. "Potential Growth of Aquatic Plants in the Lower Mekong River Basin," Washington, D.C.

61. United States Dept. of Agriculture, Soil Conservation Service, 1975. "Maintaining Watercourses," Leaflet No. 562, United States Government Printing Office, Washington, D.C.

62. United States Dept. of the Interior, Bureau of Reclamation, "Canals and Related Structures, Design Standards, No. 3," Denver, Colorado.

63. United Stated Dept. of the Interior, Bureau of Reclamation. 1975. "Concrete Manual," 8th Edition, United States Government Printing Office, Washington, D.C.

64. United States Dept. of the Interior, Bureau of Reclamation. 1978. "Drainage Manual," First Edition, United States Government Printing Office, Washington, D.C.

65. United States Dept. of the Interior, Bureau of Reclamation, 1963. "Linings for Irrigation Canals," First Edition, United States Government Printing Office, Washington, D.C.

66. United States Dept. of the Interior, Bureau of Reclamation. 1982. "Operation and Maintenance Guidelines for Small Dams," Denver, Colorado.

67. United States Dept. of the Interior, Bureau of Reclamation. 1983. "Safety Evaluation of Existing Dams," United States Government Printing Office, Denver, Colorado, USA (Revised).

68. United States Dept. of the Interior, Bureau of Reclamation, "Safety Manual, Vol. XVI, Operation and Maintenance," Denver Colorado.

69. United States Dept. of the Interior, National Park Service. 1983. "Dams and Appurtenant Works, Maintenance, Operations and Safety, Guideline NPS-40-Release No. 1," United States Government Printing Office, Washington, D.C.

70. Uppal, H.L. 1966. "Sediment Control in Rivers and Canals," Central Publication No. 79, Board of Irrigation and Power, New Delhi, India.

71. Vermeirer, L., and G.A. Jobling. 1984. "Localized Irrigation, Design, Installation, Operation, Evaluation," Paper 36, Food and Agriculture Organization of the United Nations, Rome, Italy.

72. Watts, E.J. 1962. *"Operation and Maintenance of Centrifugal Pumps,"* Journal of the American Water Works Association, New York, USA. June: 711-718.

73. Weeks, L.O., and O.J. Nordland. 1981. "Rehabilitation of Irrigation and Drainage Systems within Coachella Valley, California, U.S.A.", International Commission on Irrigation and Drainage, 11th Congress, Grenoble, Central Electric Press, New Delhi, India.

74. Zuidema, F.C., and J. Schelten. 1969. "Maintenance of Tile Drainage Systems," International Commission on Irrigation and Drainage, 7th Congress, Mexico City, Central Electric Press, New Delhi, India.

75. Plusquellec, H. 1988. "Improving the Operation of Canal Irrigation Systems," The Economic Development Institute and the Agriculture Development Department, World Bank, Washington, D.C.

76. Rural Water Commission, Victoria, Australia. 1988. "Irrigation and Drainage Practice,"

77. Clyma, W. and M. Lowdermilk. 1988. "Improving the Management of Irrigated Agriculture, A Methodology for Diagnostic Analysis," University Services Center, Colorado State University, Fort Collins, Colorado.

78. Jones, A. and W. Clyma. 1988. "Improving the Management of Irrigated Agriculture,The Management Training and Planning Program for Command Water Management, Pakistan," University Services Center, Colorado State University, Fort Collins, Colorado.

79. U.S. Agency for International Development, WASH Technical Report, No. 37, February 1988, "Guidelines for Institutional Assessment, Water and Waste Water Institutions," Washington, D.C., USA.

80. Government of Victoria, Australia. 1986. "Corporate Planning in Victorian Government, Concepts and Techniques," Department of Management and Budget, Treasury Place, Melbourne, Victoria, Australia.

81. Royal Irrigation Department, Bangkok, Thailand. 1986. "Organization, Duties and Responsibilities of Divisions, Regional Irrigation Offices, etc.," Foreign Affairs Branch, R.I.D.

82. U.N. Conference on Environment and Development; Rio de Janeiro 1992. Agenda 21, Chapter 18, Fresh Water Resources.

83. International Conference on Water and the Environment; The Dublin Statement 1992.

84. Economic Development Institute of the World Bank, Training Strategy in the Water Sector, June 1993.

85. International Commission on Irrigation and Drainage, Questionnaires on Identification of O&M Management Functions and Costs 1994 (to be published).

Distributors of World Bank Publications

Prices and credit terms vary from country to country. Consult your local distributor before placing an order.

ARGENTINA
Oficina del Libro Internacional
Av. Cordoba 1877
1120 Buenos Aires
Tel: (54 11) 815-8354
Fax: (54 11) 815-8156
E-mail: olilibro@satlink.com

AUSTRALIA, FIJI, PAPUA NEW GUINEA, SOLOMON ISLANDS, VANUATU, AND WESTERN SAMOA
D.A. Information Services
648 Whitehorse Road
Mitcham 3132
Victoria
Tel: (61) 3 9210 7777
Fax: (61) 3 9210 7788
E-mail: service@dadirect.com.au
URL: http://www.dadirect.com.au

AUSTRIA
Gerold and Co.
Weihburggasse 26
A-1011 Wien
Tel: (43 1) 512-47-31-0
Fax: (43 1) 512-47-31-29
URL: http://www.gerold.co/at.online

BANGLADESH
Micro Industries Development Assistance Society (MIDAS)
House 5, Road 16
Dhanmondi R/Area
Dhaka 1209
Tel: (880 2) 326427
Fax: (880 2) 811188

BELGIUM
Jean De Lannoy
Av. du Roi 202
1060 Brussels
Tel: (32 2) 538-5169
Fax: (32 2) 538-0841

BRAZIL
Publicações Tecnicas Internacionais Ltda.
Rua Peixoto Gomide, 209
01409 Sao Paulo, SP.
Tel: (55 11) 259-6644
Fax: (55 11) 258-6990
E-mail: postmaster@pti.uol.br
URL: http://www.uol.br

CANADA
Renouf Publishing Co. Ltd.
5369 Canotek Road
Ottawa, Ontario K1J 9J3
Tel: (613) 745-2665
Fax: (613) 745-7660
E-mail: order.dept@renoufbooks.com
URL: http://www.renoufbooks.com

CHINA
China Financial & Economic Publishing House
8, Da Fo Si Dong Jie
Beijing
Tel: (86 10) 6333-8257
Fax: (86 10) 6401-7365

China Book Import Centre
P.O. Box 2825
Beijing

COLOMBIA
Infoenlace Ltda.
Carrera 6 No. 51-21
Apartado Aereo 34270
Santafé de Bogotá, D.C.
Tel: (57 1) 285-2798
Fax: (57 1) 285-2798

COTE D'IVOIRE
Center d'Edition et de Diffusion Africaines (CEDA)
04 B.P. 541
Abidjan 04
Tel: (225) 24 6510;24 6511
Fax: (225) 25 0567

CYPRUS
Center for Applied Research
Cyprus College
6, Diogenes Street, Engomi
P.O. Box 2006
Nicosia
Tel: (357 2) 44-1730
Fax: (357 2) 46-2051

CZECH REPUBLIC
National Information Center
prodejna, Konviktská 5
CS – 113 57 Prague 1
Tel: (42 2) 2422-9433
Fax: (42 2) 2422-1484
URL: http://www.nis.cz/

DENMARK
SamfundsLitteratur
Rosenoerns Allé 11
DK-1970 Frederiksberg C
Tel: (45 31) 351942
Fax: (45 31) 357822

ECUADOR
Libri Mundi
Librería Internacional
P.O. Box 17-01-3029
Juan Leon Mera 851
Quito
Tel: (593 2) 521-606; (593 2) 544-185
Fax: (593 2) 504-209
E-mail: librimu1@librimundi.com.ec
E-mail: librimu2@librimundi.com.ec

EGYPT, ARAB REPUBLIC OF
Al Ahram Distribution Agency
Al Galaa Street
Cairo
Tel: (20 2) 578-6083
Fax: (20 2) 578-6833

The Middle East Observer
41, Sherif Street
Cairo
Tel: (20 2) 393-9732
Fax: (20 2) 393-9732

FINLAND
Akateeminen Kirjakauppa
P.O. Box 128
FIN-00101 Helsinki
Tel: (358 0) 121 4418
Fax: (358 0) 121-4435
E-mail: akatilaus@stockmann.fi
URL: http://www.akateeminen.com/

FRANCE
World Bank Publications
66, avenue d'Iéna
75116 Paris
Tel: (33 1) 40-69-30-56/57
Fax: (33 1) 40-69-30-68

GERMANY
UNO-Verlag
Poppelsdorfer Allee 55
53115 Bonn
Tel: (49 228) 949020
Fax: (49 228) 217492
URL: http://www.uno-verlag.de
E-mail: unoverlag@aol.com

GHANA
Epp Books Services
P.O. Box 44
TUC
Accra

GREECE
Papasotiriou S.A.
35, Stournara Str.
106 82 Athens
Tel: (30 1) 364-1826
Fax: (30 1) 364-8254

HAITI
Culture Diffusion
5, Rue Capois
C.P. 257
Port-au-Prince
Tel: (509) 23 9260
Fax: (509) 23 4858

HONG KONG, MACAO
Asia 2000 Ltd.
Sales & Circulation Department
Seabird House, unit 1101-02
22-28 Wyndham Street, Central
Hong Kong
Tel: (852) 2530-1409
Fax: (852) 2526-1107
E-mail: sales@asia2000.com.hk
URL: http://www.asia2000.com.hk

HUNGARY
Euro Info Service
Margitszgeti Europa Haz
H-1138 Budapest
Tel: (36 1) 350 80 24, 350 80 25
Fax: (36 1) 350 90 32
E-mail: euroinfo@mail.matav.hu

INDIA
Allied Publishers Ltd.
751 Mount Road
Madras - 600 002
Tel: (62 21) 390-4290
Fax: (62 21) 390-4289

INDONESIA
Pt. Indira Limited
Jalan Borobudur 20
P.O. Box 181
Jakarta 10320
Tel: (62 21) 390-4290
Fax: (62 21) 390-4289

IRAN
Ketab Sara Co. Publishers
Khaled Eslamboli Ave., 6th Street
Delafrooz Alley No. 8
P.O. Box 15745-733
Tehran 15117
Tel: (98 21) 8717819; 8716104
Fax: (98 21) 8712479
E-mail: ketab-sara@neda.net.ir

Kowkab Publishers
P.O. Box 19575-511
Tehran
Tel: (98 21) 258-3723
Fax: (98 21) 258-3723

IRELAND
Government Supplies Agency
Oifig an tSoláthair
4-5 Harcourt Road
Dublin 2
Tel: (353 1) 661-3111
Fax: (353 1) 475-2670

ISRAEL
Yozmot Literature Ltd.
P.O. Box 56055
3 Yohanan Hasandlar Street
Tel Aviv 61560
Tel: (972 3) 5285-397
Fax: (972 3) 5285-397

Palestinian Authority/Middle East
Index Information Services
P.O.B. 19502 Jerusalem
Tel: (972 2) 6271219
Fax: (972 2) 6271634

ITALY
Licosa Commissionaria Sansoni SPA
Via Duca Di Calabria, 1/1
Casella Postale 552
50125 Firenze
Tel: (55) 645-415
Fax: (55) 641-257
E-mail: licosa@ftbcc.it
URL: http://www.ftbcc.it/licosa

JAMAICA
Ian Randle Publishers Ltd.
206 Old Hope Road, Kingston 6
Tel: 876-927-2085
Fax: 876-977-0243
E-mail: irpl@colis.com

JAPAN
Eastern Book Service
3-13 Hongo 3-chome, Bunkyo-ku
Tokyo 113
Tel: (81 3) 3818-0861
Fax: (81 3) 3818-0864
E-mail: orders@svt-ebs.co.jp
URL: http://www.bekkoame.or.jp/~svt-ebs

KENYA
Africa Book Service (E.A.) Ltd.
Quaran House, Mfangano Street
P.O. Box 45245
Nairobi
Tel: (254 2) 223 641
Fax: (254 2) 330 272

KOREA, REPUBLIC OF
Deejon Trading Co. Ltd.
P.O. Box 34, Youida, 706 Seoun Bldg
44-6 Youido-Dong, Yeongchengo-Ku
Seoul
Tel: (82 2) 785-1631/4
Fax: (82 2) 784-0315

MALAYSIA
University of Malaya Cooperative Bookshop, Limited
P.O. Box 1127
Jalan Pantai Baru
59700 Kuala Lumpur
Tel: (60 3) 756-5000
Fax: (60 3) 755-4424
E-mail: umkoop@tm.net.my

MEXICO
INFOTEC
Av. San Fernando No. 37
Col. Toriello Guerra
14050 Mexico, D.F.
Tel: (52 5) 624-2800
Fax: (52 5) 624-2822
E-mail: infotec@rtn.net.mx
URL: http://rtn.net.mx

Mundi-Prensa Mexico S.A. de C.V.
c/Rio Panuco, 141-Colonia Cuauhtemoc
06500 Mexico, D.F.
Tel: (52 5) 533-5658
Fax: (52 5) 514-6799

NEPAL
Everest Media International Services (P) Ltd.
GPO Box 5443
Kathmandu
Tel: (977 1) 472 152
Fax: (977 1) 224 431

NETHERLANDS
De Lindeboom/InOr-Publikaties
P.O. Box 202, 7480 AE Haaksbergen
Tel: (31 53) 574-0004
Fax: (31 53) 572-9296
E-mail: lindeboo@worldonline.nl
URL: http://www.worldonline.nl/~lindeboo

NEW ZEALAND
EBSCO NZ Ltd.
Private Mail Bag 99914
New Market
Auckland
Tel: (64 9) 524-8119
Fax: (64 9) 524-8067

NIGERIA
University Press Limited
Three Crowns Building Jericho
Private Mail Bag 5095
Ibadan
Tel: (234 22) 41-1356
Fax: (234 22) 41-2056

NORWAY
NIC Info A/S
Book Department, Postboks 6512 Etterstad
N-0606 Oslo
Tel: (47 22) 97-4500
Fax: (47 22) 97-4545

PAKISTAN
Mirza Book Agency
65, Shahrah-e-Quaid-e-Azam
Lahore 54000
Tel: (92 42) 735 3601
Fax: (92 42) 576 3714

Oxford University Press
5 Bangalore Town
Sharae Faisal
P.O. Box 13033
Karachi-75350
Tel: (92 21) 446307
Fax: (92 21) 4547640
E-mail: ouppak@TheOffice.net

Pak Book Corporation
Aziz Chambers 21, Queen's Road
Lahore
Tel: (92 42) 636 3222; 636 0885
Fax: (92 42) 636 2328
E-mail: pbc@brain.net.pk

PERU
Editorial Desarrollo SA
Apartado 3824, Lima 1
Tel: (51 14) 285380
Fax: (51 14) 286628

PHILIPPINES
International Booksource Center Inc.
1127-A Antipolo St, Barangay, Venezuela
Makati City
Tel: (63 2) 896 6501; 6505; 6507
Fax: (63 2) 896 1741

POLAND
International Publishing Service
Ul. Piekna 31/37
00-677 Warzawa
Tel: (48 2) 628-6089
Fax: (48 2) 621-7255
E-mail: books%ips@ikp.atm.com.pl
URL: http://www.ipscg.waw.pl/ips/export/

PORTUGAL
Livraria Portugal
Apartado 2681, Rua Do Carmo 70-74
1200 Lisbon
Tel: (1) 347-4982
Fax: (1) 347-0264

ROMANIA
Compani De Librari Bucuresti S.A.
Str. Lipscani no. 26, sector 3
Bucharest
Tel: (40 1) 613 9645
Fax: (40 1) 312 4000

RUSSIAN FEDERATION
Isdatelstvo <Ves Mir>
9a, Kolpachniy Pereulok
Moscow 101831
Tel: (7 095) 917 87 49
Fax: (7 095) 917 92 59

SINGAPORE, TAIWAN, MYANMAR, BRUNEI
Ashgate Publishing Asia Pacific Pte. Ltd.
41 Kallang Pudding Road #04-03
Golden Wheel Building
Singapore 349316
Tel: (65) 741-5166
Fax: (65) 742-9356
E-mail: ashgate@asianconnect.com

SLOVENIA
Gospodarski Vestnik Publishing Group
Dunajska cesta 5
1000 Ljubljana
Tel: (386 61) 133 83 47; 132 12 30
Fax: (386 61) 133 80 30
E-mail: repansekj@gvestnik.si

SOUTH AFRICA, BOTSWANA
For single titles:
Oxford University Press Southern Africa
Vasco Boulevard, Goodwood
P.O. Box 12119, N1 City 7463
Cape Town
Tel: (27 21) 595 4400
Fax: (27 21) 595 4430
E-mail: oxford@oup.co.za

For subscription orders:
International Subscription Service
P.O. Box 41095
Craighall
Johannesburg 2024
Tel: (27 11) 880-1448
Fax: (27 11) 880-6248
E-mail: iss@is.co.za

SPAIN
Mundi-Prensa Libros, S.A.
Castello 37
28001 Madrid
Tel: (34 1) 431-3399
Fax: (34 1) 575-3998
E-mail: libreria@mundiprensa.es
URL: http://www.mundiprensa.es/

Mundi-Prensa Barcelona
Consell de Cent, 391
08009 Barcelona
Tel: (34 3) 488-3492
Fax: (34 3) 487-7659
E-mail: barcelona@mundiprensa.es

SRI LANKA, THE MALDIVES
Lake House Bookshop
100, Sir Chittampalam Gardiner Mawatha
Colombo 2
Tel: (94 1) 32105
Fax: (94 1) 432104
E-mail: LHL@sri.lanka.net

SWEDEN
Wennergren-Williams AB
P.O. Box 1305
S-171 25 Solna
Tel: (46 8) 705-97-50
Fax: (46 8) 27-00-71
E-mail: mail@wwi.se

SWITZERLAND
Librairie Payot Service Institutionnel
Côtes-de-Montbenon 30
1002 Lausanne
Tel: (41 21) 341-3229
Fax: (41 21) 341-3235

ADECO Van Diermen EditionsTechniques
Ch. de Lacuez 41
CH1807 Blonay
Tel: (41 21) 943 2673
Fax: (41 21) 943 3605

THAILAND
Central Books Distribution
306 Silom Road
Bangkok 10500
Tel: (66 2) 235-5400
Fax: (66 2) 237-8321

TRINIDAD & TOBAGO AND THE CARRIBBEAN
Systematics Studies Ltd.
St. Augustine Shopping Center
Eastern Main Road, St. Augustine
Trinidad & Tobago, West Indies
Tel: (868) 645-8466
Fax: (868) 645-8467
E-mail: tobe@trinidad.net

UGANDA
Gustro Ltd.
PO Box 9997, Madhvani Building
Plot 16/4 Jinja Rd.
Kampala
Tel: (256 41) 251 467
Fax: (256 41) 251 468
E-mail: gus@swiftuganda.com

UNITED KINGDOM
Microinfo Ltd.
P.O. Box 3, Alton, Hampshire GU34 2PG
England
Tel: (44 1420) 86848
Fax: (44 1420) 89889
E-mail: wbank@ukminfo.demon.co.uk
URL: http://www.microinfo.co.uk

VENEZUELA
Tecni-Ciencia Libros, S.A.
Centro Cuidad Comercial Tamanco
Nivel C2, Caracas
Tel: (58 2) 959 5547; 5035; 0016
Fax: (58 2) 959 5636

ZAMBIA
University Bookshop, University of Zambia
Great East Road Campus
P.O. Box 32379
Lusaka
Tel: (260 1) 252 576
Fax: (260 1) 253 952